ОПЫТЪ СИСТЕМЫ ЭЛЕМЕНТОВЪ.

ОСНОВАННОЙ НА ИХЪ АТОМНОМЪ ВѢСѢ И ХИМИЧЕСКОМЪ СХОДСТВѢ.

		Ti = 50	Zr = 90	? = 180.
		V = 51	Nb = 94	Ta = 182.
		Cr = 52	Mo = 96	W = 186.
		Mn = 55	Rh = 104,4	Pt = 197,4.
		Fe = 56	Rn = 104,4	Ir = 198.
		Ni = Co = 59	Pl = 106,6	Os = 199.
H = 1		Cu = 63,4	Ag = 108	Hg = 200.
	Be = 9,4	Mg = 24	Zn = 65,2	Cd = 112
	B = 11	Al = 27,4	? = 68	Ur = 116	Au = 197?
	C = 12	Si = 28	? = 70	Sn = 118
	N = 14	P = 31	As = 75	Sb = 122	Bi = 210?
	O = 16	S = 32	Se = 79,4	Te = 128?
	F = 19	Cl = 35,5	Br = 80	I = 127
Li = 7	Na = 23	K = 39	Rb = 85,4	Cs = 133	Tl = 204.
		Ca = 40	Sr = 87,6	Ba = 137	Pb = 207.
		? = 45	Ce = 92
		?Er = 56	La = 94
		?Yt = 60	Di = 95
		?In = 75,6	Th = 118?

Д. Менделѣевъ

DECODING THE PERIODIC TABLE

Jurjen van der Wal

iUniverse, Inc.
Bloomington

Decoding the Periodic Table

iUniverse books may be ordered through booksellers or by contacting:

iUniverse
1663 Liberty Drive
Bloomington, IN 47403
www.iuniverse.com
1-800-Authors (1-800-288-4677)

ISBN: 978-1-4401-8672-1 (sc)
ISBN: 978-1-4401-8673-8 (e)

Printed in the United States of America

iUniverse rev. date: 5/19/2011

DECODING
THE PERIODIC TABLE

INTRODUCTION

During my days in college in the Netherlands I was always intrigued by the large chart with the Periodic Table of Elements that hung on the wall of the Chemistry Room.

On the very last day of school I asked my teacher if there was some sort of particular physical structure of the assembly of all of those elements together.

In response she said that she did not know what that structure was, but then with chalk she wrote on the blackboard the number of elements in each period:

Period 1: $2x1^2=2$, Period 2::$2x2^2=8$, Period 3: $2x2^2=8$
Period 4: $2x3^2-18$, Period 5: $2x3^2-18$, Period 6: $2x4^2=32$.

It was that quadratic relationship that really struck me. I never forgot that. Decades later I experimented on paper with various approaches such as concentric spheres in the late 1960's that one way or the other might have quadratic relationships. But that whole effort got nowhere.

In 1989 I started yet another effort when, like a bolt of lightening out of the sky, my mind got the notion that nature's particles such as proton, neutron, etc. were made with straight lines only and there would be no curvatures anywhere, and that would mean that there would be no concentric spheres. It was a revolutionary thought, to say the least. Had anyone ever considered that there would be straight lines only, and no curves of any kind?

What made me take it serious was the conviction that the Physical Sciences around the world seemed to be getting absolutely nowhere with their all out attempts to put several new theories together, and even now they are still not sure if that 8 billion dollar new Large Hadron Collider on the Swiss/French border will produce the hoped-for results.

I have worked all alone on this, it was a great joy to work on it, sometimes feeing that I was all by myself far out there in the universe, asking the protons and neutrons in my brain to tell me what they looked like, and eventually, I think, they did. The result is this book.

Date: March 2010

By personally not being involved with any physical sciences institute that does research in Particle Physics, it was often not easy to find helpful information that might be useful in this quest to formulate a new and very unconventional theory of particle physics.

However, this dearth of information actually was an advantage, because the lack of a history in this novel straight-line approach towards a new theory would not be influenced by some else's efforts that might have contaminated my mind as it entered the territory of this new approach. In the meantime I stayed in touch with the latest development in physics by subscribing to publications such as Scientific American, Science News, Discover, SEED, American Scientist, etc., which have found many eager subscribers in the general public which are fascinated by the as yet unsolved riddles of this universe that is our home.

This theory owes everything to Sir Isaac Newton, 1642-1727, Scientist, who in 1686 published "Principia Mathematica" with the following Laws:

1. An object that is moving steadily will remain in that same state till a force is applied to it.

2. The acceleration of an object depends on two things: the object's mass and the force acting upon it.

3. For every action there is an equal and opposite reaction.

In this theory:

Newton's 3rd Law is everywhere.

Not a day seems to go by without one way or the other reading in a newspaper or seeing on TV something about energy, dark matter, the big bang, black holes, as well as the search for elusive particles, everywhere around the world.

What we are not seeing is that we don't seem to get any closer to uncover any of those many unknowns in physics that control all facets of our lives. Billions of dollars and euros are being spent, and we are getting nowhere.

This tells me that something is drastically wrong with the approach that scientists have employed to find their evasive targets.

What I do know for sure is that the Periodic Table of Elements can be seen as a combination of the Rosetta Stone and the hieroglyphs of Egypt, which in this case needs to be dissected and analyzed as to why it has those sequences of its quadratic and therefore square relationships in its periods. What does that actually tell us, and how can we exploit that?

This theory has an answer to that with "Decoding the Periodic Table," which in my opinion fits within all of the parameters that it needs to comply with in order to ensure its validity.

DECODING
THE PERIODIC TABLE

―――――――――― **ABSTRACT** ――――――――――

For the past 100 years countless efforts by physicists to find a theory that would explain all of the interactions between chemicals as well as their transformations into other forms have failed. What this continuing failure means is that somehow, somewhere something must be drastically wrong with the method and/or the ground rules that are being deployed to attack the problem.

A _ _ In this theory the make-up of the Periodic Table is the basis for all that is physics-related such as the actual shape of particles like the neutron, proton, electron and neutrino.

> The Periodic Table's series of element quantities show a quadratic relationship:
> Period 1, from Hydrogen to Helium   , has 2, or 2×1^2 elements.
> Period 2, from Helium to Neon   , has 8, or 2×2^2 elements.
> Period 3, from Neon to Argon   , has 8, or 2×2^2 elements.
> Period 4, from Argon to Krypton   , has 18, or 2×3^2 elements.
> Period 5, from Krypton to Xenon   , has 18, or 2×3^2 elements.
> Period 6, from Xenon to Radon   , has 32, or 2×4^2 elements.
> Period 7, from Radon to 'ZZZ' ???   , (has 32, or 2×4^2 elements).

In support of the above: There is a 90° angle between the electric and magnetic components of an electromagnetic wave, such as light.

These quadratic expressions mean that there actually might be particles with square structures and straight lines, but no spheres or curved blobs, nor those wiggly rings of the string theory. This straight line idea could possibly be applied to Particle Physics, and why not, everything else had failed.

B _ _ This is a Theory of Architectural Design of which its basic building block is a square that serves as the common base for 2 pyramids with 45 degree slopes, one above and one below it. Those pyramids consist of an assembly of small cubes which represent a force that is perpendicular to one of the cube's sides. These cube's forces at each of the double pyramid's four sides point in four different directions, North, South, East, West, as an easy marker, and no, it does not mean magnetism, just direction. These cubes are connected to each other at their corners, and are separated by vacant cubic spaces. The overall effect of such a double pyramid assembly of forces equals zero.

C _ _ The next step is to put the square base of one of those double pyramids down in a "horizontal" position, and then make a "plus" sign by attaching in a horizontal manner an identical double pyramid square to its North, East, South and West side. Next: Fold up the plus sign's North and West attached double pyramids UP till their sloped sides come to rest on the original center pyramids, and then do the same at the center pyramid's Southeast corner, by folding those double pyramids DOWN. We now have created a NEUTRON, as will be shown here.

D _ _ A free neutron will fall apart in 15 minutes into a proton and electron which both will lose part of their structure to make a neutrino. The forces of the small cubes within these particles are intrinsic, and, subject to the location of those cubes within the overall structure they are in, they will be observed as being gravity, strong force, weak force or magnetism. It all fits, including being neutral, positive or negative, quarks, and the generation of nuclear mass which involved Newton's 3rd Law about action/reaction forces.

E _ _ For this theory, because of its actual involvement of what its particles' structure might look like, it became useful to convert their measures nuclear mass into a system of which its electron's nuclear mass would be 1,0000.

Measured Nuclear Mass		Converted Nuclear Mass	
Electron	.511 MeV	Electron	1.0000
Neutron	939.5732 MeV	Neutron	1838.6949
Proton	938.2792 MeV	Proton	1836.1636

Having tried for many years to find a nice and clean nuclear mass structure that would make sense inside a pyramid shape, the answer came when I tried a cubic volume for the neutron's converted nuclear mass of 1838.6949, and found $12.25^3 = 1838.2656$, nearly a perfect hit. However: $12.25 = (3½)^2$, which changes 12.25^3 to $(3½)^6$, $= 1838.2656$ (mass). Using that number as the standard value is it relates to other particles:

The new proton's measure mass = 1835.7344
The new electron's measured mass = 0.5111198 MeV.

The interesting thing is that the fractional portion of the neutron's mass of 1838.2656 dovetails exactly with the proton's mass of 1835.7344, so that their combination is 1838.2656 + 1835.7344 = 3674.0000. Helium mass is then twice that, = 7348,000. This 'clean' whole number is ideal for Helium which is the building block for the entire Periodic Table.

This perfect dovetailing reminds one of biology's double helix where fulfillment is achieved when Group A merges with Group T and Group C merges with Group G.

The neutron's mass number $(3½)^6$ equals $(3½)^3$ x $(3½)^3$, which means there are 2 identical opposing cubes packed with forces which are complying with Newton's 3rd Law, Action = Reaction.

F _ _
- This theory has no concentric spheres, but it has cubes which for the complete Periodic Table are stacked in 2 layers of 3x3 = 9 cubes, for a total of 18 cubes. The total assembly has 75 cube squares, each of which is suitable to receive the square double pyramid base of Helium, with their 45 degree pyramid slopes perfectly merging with their neighbors.

- This compact assembly starts with Period 1, Helium, then comes Period 2 with 4 Heliums joining in, making Neon.

- Then Period 3 with 4 more Heliums comes in, making Argon by completing a cube above and also a cube below the original Helium, for a total of 9 Heliums, with its 18 protons, making Argon's 18 elements.

- This Argon structure is like a pod, and it is the major sub-assembly that nature will use again to make Krypton, Xenon and Radon.

- The 14-unit Lanthanium and Actinium groups serve as upper and lower covers for the whole assembly.

CHEMICAL
Periodic Table

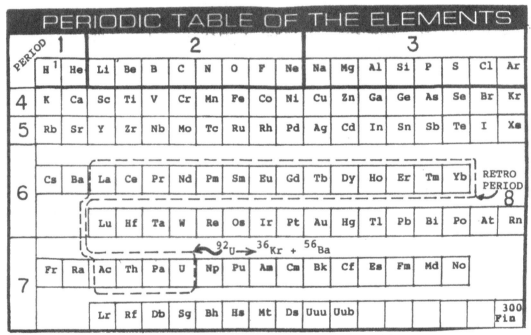

STRUCTURAL

PERIODIC TABLE OF THE ELEMENTS

The Periodic Table's series of element quantities show a quadratic relationship:

Period 1, from Hydrogen to Helium , has 2, or 2×1^2 elements.
Period 2, from Helium to Neon , has 8, or 2×2^2 elements.
Period 3, from Neon to Argon , has 8, or 2×2^2 elements.
Period 4, from Argon to Krypton , has 18, or 2×3^2 elements.
Period 5, from Krypton to Xenon , has 18, or 2×3^2 elements.
Period 6, from Xenon to Radon , has 32, or 2×4^2 elements.
Period 7, from Radon to 'ZZZ' ??? , (has 32, or 2×4^2 elements).

6

TABLE OF CONTENTS

It was now 1989.

Having 'tried several unsuccessful extensive efforts in the past to solve (some) of this problem, this was a new approach.

On the premise that the void is capable of containing points, lines and areas, we can then propose the image of Figure 1, in which an infinitesimally small area of the pure void has been selected for further scrutiny. Because of earlier listed examples of quadratic relationships in the Periodic Table and the rectangular offset between the electric and magnetic wave components in an electromagnetic wave, there are reasonable grounds to draw a matrix-like pattern that consists of two sets of parallel lines that intersect each other so that a large number of identical small squares will fill a square area. Vanishing points R and S are shown as a reference to the infinite range of the void.

Because of the void being an empty space by definition, we would theoretically not violate that emptiness if we would alternatingly assign plus and minus signs to those intersecting lines, because the algebraic sum of the entire assembly would then still be equal to zero.

At the nuclear level of particles there are indeed positive and negative charges, but because there are four known nuclear forces, the decision was made at this point not to use plus and minus signs but to use alternating arrows that represent some as yet unidentified force(s). Because there are four nuclear forces it was decided to try an arrangement of two groups of four intersecting forces in the matrix, as shown on the next page.

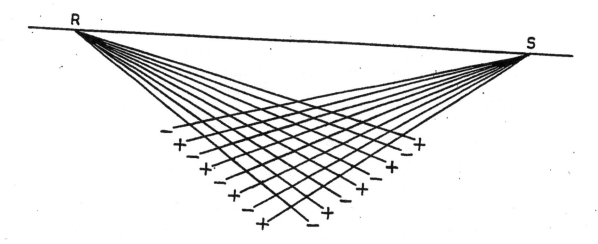

Fig, 1

Fig. 2 displays the plus and minus concept of the subdividing lines, which in Fig. 3 are replaced by a system of alternating flow lines. In this manner we are dealing with nine small squares which are a part of a larger square. The immediate nice aspect of this arrangement was the fact that there is one square located exactly in the center of the other squares, and this gave rise to the expectation that this center square might be the home of nuclear mass, thereby assuming that the nine-square-assembly would one way or the other represent a neutron or a proton.

All eight intersecting lines have been assigned alternating flow directions, and each of the nine squares has been numbered.

The task was now to evaluate this matrix and hopefully find something that would indicate that this concept had potential. Having been influenced considerably by reading and hearing about the curvature of space-time I originally coiled the matrix along the 1-5-9 axis so that square 3 would overlap square 7.

To be frank: I got nowhere with it. It was 1991.

I never gave up on this concept of a matrix in the void, it was such a simple and straight-forward concept.

It is said that one cannot unring a bell, and in the summer of 1998 I was again drawn towards the void and its matrix when I finally realized that this flat matrix had no reason to curve at all, because there was no force acting in any kind of direction that had a component that was perpendicular to the matrix's plane. The straight lines had to remain straight at all times, with no exceptions. And after that realization: Everything began to happen, slowly.

Knowing that a single neutron, by itself, will split into a proton, an electron and a neutrino was a big help in this pursuit to find specific particle structures.

It was on that premise that the basic pattern of Fig. 3 might represent the neutron, with its nuclear mass in its center square, and that its break-up might indeed produce the sought for proton, electron and neutrino.

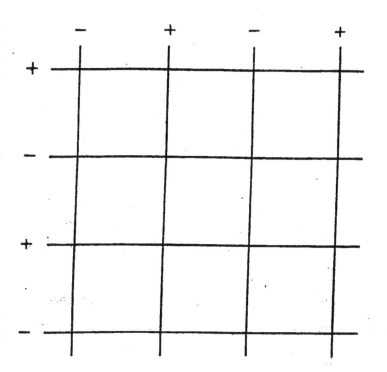

Fig. 2

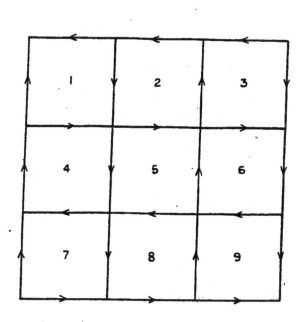

Fig. 3

Ever since conceiving in 1989 this concept of using the rectangular matrix of the previous page, I tried various approaches in attempts to find unknown common denominators which would provide a bridge between particles like the neutron, proton electron and neutrino, as well as gravity, radiation, etc.

This entire effort was based on the physical geometric shapes of those particles, whereby efforts were being made to identify the locations or specific sites in these structures that might be harboring the gravitational pull, electric charge, etc.

Up to this far, the nine squares in Fig. 3 only represented a flat 2-dimensional pattern with zero thickness. This search however was for 3-dimensional particles, and therefore an educated guess had to be made as to how to give a 3-D shape to this nascent pattern.

The choice fell on square pyramid shapes for each square, with two bottom to bottom connected pyramids sharing a common base plate.

In addition it was decided to assign a 45° slope to the angle that each pyramid's side slope makes with its pyramid's baseplate. The reason for the 45° slope was the thought that the particles which hopefully somehow might be found in these structures could easily and snugly connect with each other if and when 45° slopes would fold over and produce a 90° angle between their two respective baseplates.

What I had been doing was the fact that I had used all nine squares of the matrix to find the solution to this quest. After years of trial and error, the nine pyramids turned eventually out to be an excessive number, only five of the nine bottom to bottom pyramid pairs were needed. Ockham's razor did the trick: Squares 1, 3, 7 and 9 were eliminated, leaving only those squares that make a cross or plus-sign: 2, 4, 5, 6 and 8. I left their numbering as it originally was, as a reference to their original position, and also because it is an easy pattern to remember, because this pattern is also on our telephone pad.

Fig. 4 shows the original nine-square pyramid pattern, which did not work, and Fig. 5 is the 5-square pattern that eventually would lead to the solution of this endeavor.

Again, for easy identification purposes, the slopes of the pyramids in Fig. 5 were assigned the directions of a compass, North, East, South and West. In this situation there is a top pyramid with a bottom pyramid directly below it. An example: the southern slope of the upper pyramid in square 6 in Fig. 5a is identified as 6S, and the corresponding pyramid slope of the bottom pyramid directly below it is marked with 6SO, in which the 'O' stands for Opposite.

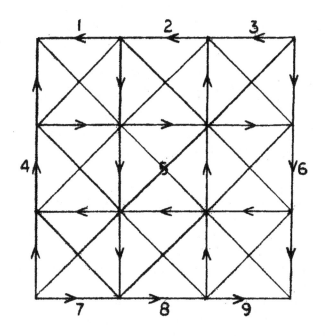

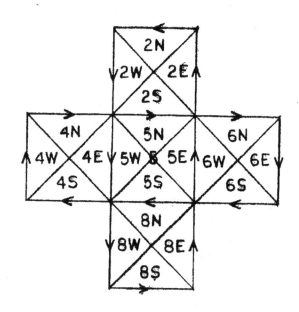

Fig. 4   Fig. 5

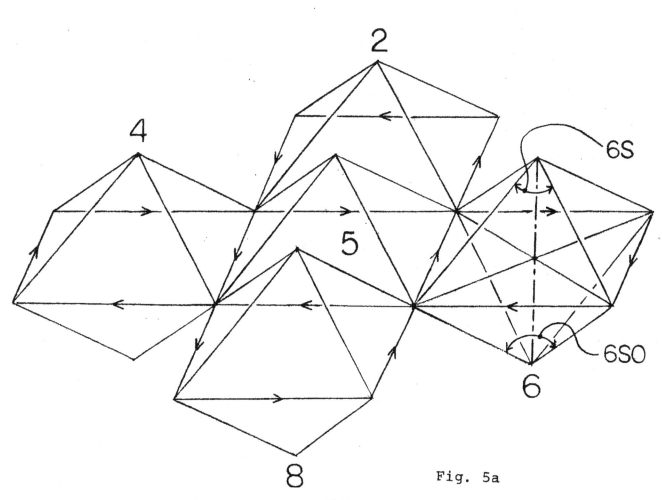

Fig. 5a

13

Fig. 6 indicates how all of the arrows of squares 2, 4, 6 and 8 make a unique pattern of CW and CCW rotations.

Central section 5 is an important exception to that, next to its vacant NW and SE corners are pyramid baselines with two diverging arrows, its NE and SW corners are next to baselines with two converging arrows. These arrow flows are in this theory are the key to everything else that will happen here, it turned out that here in Fig. 6 you are seeing the plan view of the entire one and only basic law of Physics in this theory that controls all events in nature. More details will be shown later.

Early on it was concluded that, in order to have a versatile pyramid system, that the pyramid slopes should have a 45° angle with their base, Fig. 7a. The pyramids' other requirement, in Fig. 7b is to have a pyramid base that has the same length as the pyramids' centerpost. This is a crucial characteristic of this system. Without this kind of relationship this system would not work. The important feature here is that the baseline with the centerpost makes a tetrahedron.

Following that path of making an educated guess, we can use the knowledge that an electric current through a conductor is surrounded by a magnetic field that has a CW direction around the conductor. Fig. 7a,b both satisfy this electric current scheme where both the baseline and the arrowed centerpost qualify as examples.

HOWEVER: THIS CENTERLINE MARKER ARROW IS ONLY A SUGGESTIVE MARKER, IT DOES NOT MEAN THAT THERE IS A FORCE OR SPEED OR WHATEVER THAT GOES IN THAT DIRECTION, IT IS ONLY SHOWN AS A TRIAL BALLOON FOR MAINLY AN ELECTRIC/MAGNETIC RELATIONSHIP.

The pyramids of Fig. 8a and 8b represent section 5 with its diverging/converging arrows. Its baseline (circled) arrows are shown with their appropriate CW/CCW notation in reference to their centerposts' arrow. Except for the different directions of the baselines' arrows and the different locations where they are, their overall general symmetry provides for future exchangeabilities between such identical structures. Structurally, all pyramids have exactly the same shape and size.

Additionally, because the pyramid slopes of those pyramids make a 45° angle with the bottom to bottom pyramids' common base, this will become important when, as an example, section 2 with its pyramids folds upward 90°, so that its 2S pyramid slope comes to rest on central section 5's pyramid slope 5N.

Sections 4, 6 and 8 can do the same, either up or down. If for instance sections 2, 4, 5 and 8 would all fold up, then, respectively, their baselines of 2N, 4W, 6E and 8S would all connect on top, making the cube of Fig. 9.

We have now the beginning of a comprehensive 3-D spatial matrix.

Fig. 10 shows the geometric length relationship of every basic dimension of this pyramid system.

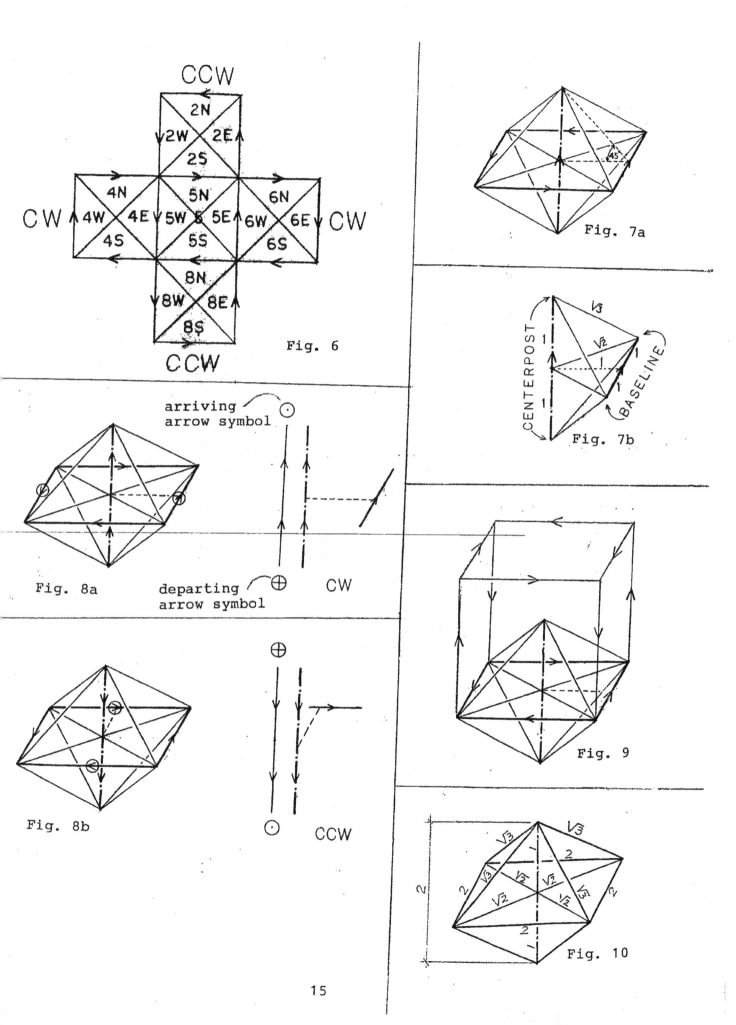

CCW

2N

2W 2E

2S

CW 4N 5N 6N CW

4W 4E 5W & 5E 6W 6E

4S 5S 6S

8N

8W 8E

8S

CCW

Fig. 6

Fig. 7a

√3

CENTERPOST √2

1 1

1 BASELINE

1

Fig. 7b

arriving
arrow symbol ⊙

Fig. 8a

departing ⊕ CW
arrow symbol

⊕

Fig. 8b

⊙ CCW

Fig. 9

√3 √3

√3 1 2

2 √2 √2

√2 √3 2

2 1

Fig. 10

15

Much has been written about symmetry and supersymmetry.

In this theory the symmetry issue permeates the entire system, and when that is applied to this pyramid system of Fig. 5 and 6, then it becomes a rather intriguing situation.

Figure 11a indicates how the directions of the centerpost arrows are (according to Fig. 7 and 8) related to the rotational direction of the pyramid's baselines, which is clockwise, CW.

This figure will be used here as the 'standard,' the ORIGINAL.

Fig. 11b is its mirror image.

The centerpost arrow's relationship with the baselines' arrows has now become CCW.

Figure 12a is like Fig. 11a, it is the original set up, and next to it is Fig. 12b, which shows the original drawing after a 90° rotation, which may be either CW or CCW, it does not matter.

The result is here, as an example, that pyramid 4 has taken the place of pyramid 2, which has as a result that the original CCW rotation of that place in space has been replaced by a CW rotation.

In this new arrangement the CW relationship between all arrows in the centerposts and their respective baselines remain unchanged.

In the process of developing this theory it became clear that nature's entire spatial relationships in particle physics might be controlled by a precise interactive system of a variety of the 4 known different forces along the direction of the system's arrows.

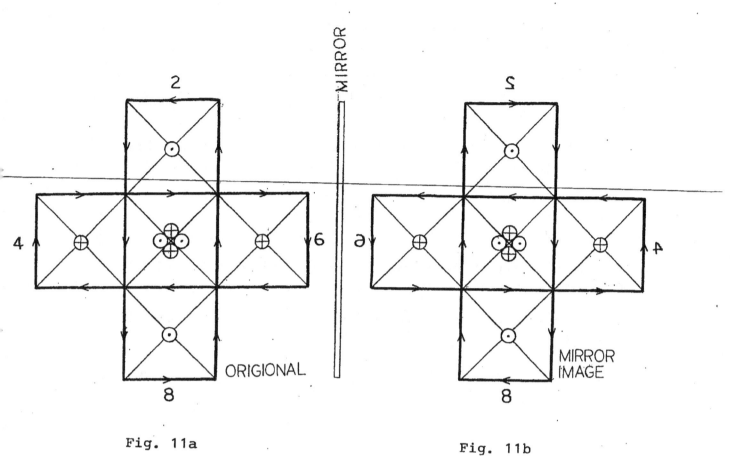

Fig. 11a

Fig. 11b

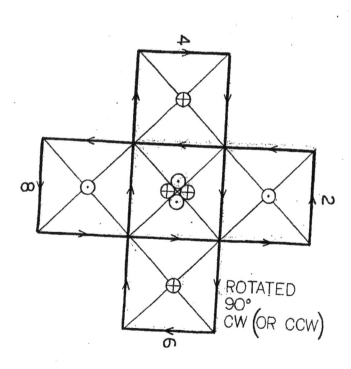

Fig. 12a

Fig. 12b

When I originally hit upon the square pyramid idea, the thought of how to package and/or assemble them with other cubes like that directly pointed to the solution of giving the pyramid slopes a 45° angle with their baseplate, see Fig. 7a.

It seemed to be the right thing to do, it was the practical thing to do, nature is precise and tidy, there never are loose ends, in an interaction between particles the outcome is always precisely the same. In this theory that notion was put into practice throughout its entire development, it has great consistency, and above all, it works.

Fig. 13 displays the way a cube can be subdivided into six identical pyramids, with no leftovers.

An identical single pyramid has been added to the bottom of the cube, where it is then via its baseplate merged with the cube's lowest inner pyramid.

By doing so, we have made a bottom to bottom pyramid pair that was already shown in Fig. 5a, 7ab, 8ab, 9 and 10.

In conjunction with another similar system that has not yet been shown here, its extended arrow pattern will become the matrix of the structure of particles.

Figure 14 was already shown as Fig. 5. Its identification markers will help the reader and researcher to follow the activities of this reverse engineering activity.

What will happen from here on out is that the four outlying pyramids, being section 2, 4, 6 and 8, will start to fold over onto central section 5, folding either up or down.

If all four sections, 2, 4, 6 and 8, would fold upward or downward, then their baselines would make a cube's configuration, as in Fig. 9 and 13.

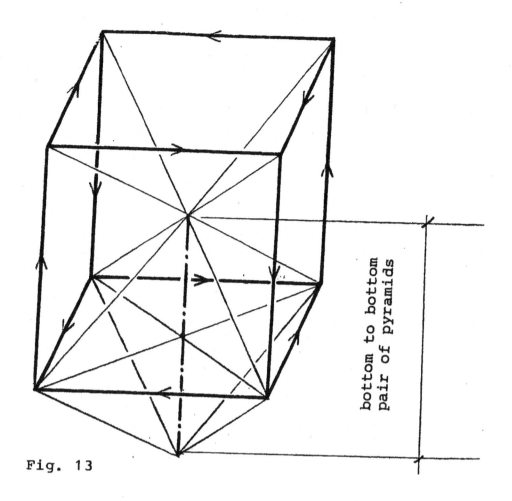

Fig. 13

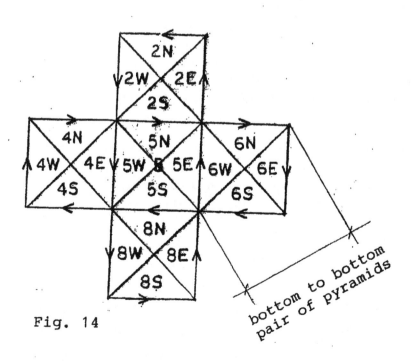

Fig. 14

From the very beginning the premise of this theory was that this initial group of double pyramids might be representing the basic components of the neutron, and that this particular assembly somehow would break up in pieces that would from then on exist as the proton, the electron and the neutrino.

Let's give it a try right here.

By splitting up the cross of the five pyramid pairs as shown in Fig. 15, which was supposed to be the neutron, I created some sort of a racetrack that would then supposedly contain the components on the outside which would assemble into an electron, while the remaining central portion would remain behind as the centerpiece, the proton.

[In earlier, unsuccessful efforts, I had been using the now discarded squares 1, 3, 7 and 9 to make the neutrino, but that misconception after a decade of fruitless efforts eventually was abandoned for an arrangement that was quite effective, in which both the proton and electron give up a small part of themselves, which then assemble into a neutrino. More about that later.]

The 4 circled arrows of the pattern seemed to suggest strongly that they represented the inward pulling force of gravity, thus hinting that this was the proton. It turned out much later that this idea was correct.

The eight 'plus' signs as shown on the proton were just a guess, which eventually turned out to be exactly on the spot where the proton has its positive electric charge, that is at square 5's upper left and lower right.

The negative components at the outside of the proton will be shown later to merge into a single structure which is the electron.

The racetrack's oval slant in Fig. 15 goes from the upper left to the lower right of the proton, it somewhat simulates or suggests that this might be the orbit that the electron later will occupy. But this will only happen after the four separate pieces of the electron that are shown here have assembled themselves in a new particle that will display all of those characteristics that the electron is supposed to possess.

More about that later in these pages.

All along it was assumed that the nuclear mass of the neutron as well as that of the proton would be located within square 5, as it turned out to be the case. This will be shown later on too.

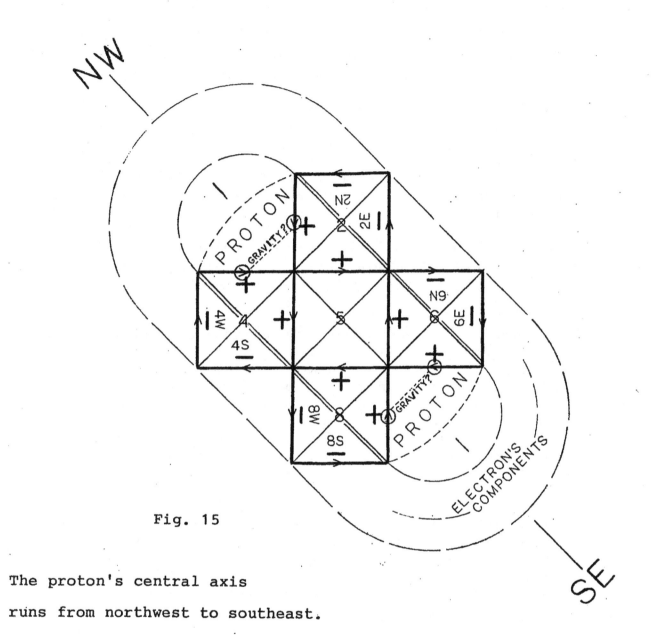

Fig. 15

The proton's central axis

runs from northwest to southeast.

21

Fig. 16a identifies all of the pyramid slopes that can be seen from above.

The pyramid slopes that are pointing downward to the backside of this paper have the same markings as those seen from above, except they have the capital letter 'O' added to their marker, so that for instance 4E becomes 4EO, etc., as shown earlier in Fig. 5a.

Fig. 16b shows the proton all by itself, all of its pyramid slopes have been given positive signs, parallel to the NW/SE diagonal.

Fig. 17a shows the whole neutron again, and when for the sake of making a comparison with the proton's configuration we single out a proton-like shape of the neutron in the NE/SW direction, then we get Fig. 17b. This is the core of the neutron, there is balance between the positive and negative signs, which make the neutron a neutral particle.

In Fig. 16b the circled arrows of Fig. 15 are shown again, their inward flow suggesting that these are the pull of gravity, and in the neutron's drawing of Fig. 17b the outward flows of the arrows along the NE/SW diagonal which have a square around them will later be shown to represent the weak force.

In these drawings this weak force is not shown in the proton drawing of Fig 16b, even though it is known that the proton has a weak force. The reason why the proton still has a weak force anyway is that each of the weak force components of the neutron is actually a bodypart of each of the four ejected pieces of the electron that was to be assembled.

A proton is basically a damaged neutron that had lost its limbs. Being immersed in the yet to be shown lattice of the matrix of its surrounding space, the proton will now attempt to regrow its lost electron parts, not unlike the split twisted ladder of biology's double helix that will start to repair itself.

The proton does not go as far in rebuilding as the double helix does, but it will surround itself with a rudimentary lattice of the lost electron pieces, and this lattice will be a soft version of the weak force. Subject weak force lattice locations at the NE and SW sites of the Fig. 17b's neutron are not shown in Fig. 16b of the proton, even though that they are there, as will be shown later in more detailed explanations.

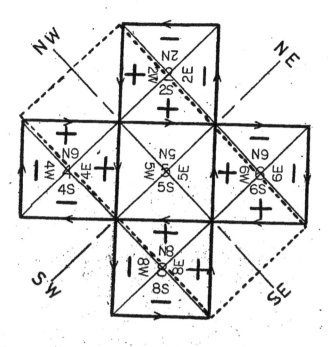

Fig. 16a

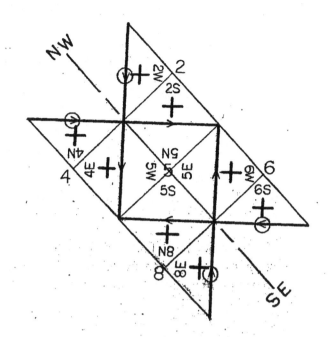

Fig. 16b

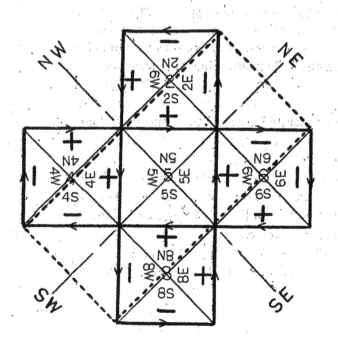

Fig. 17a

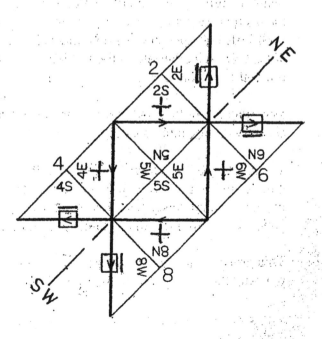

Fig. 17b

23

If there is one thing that really became apparent during this analysis of the shape of particles it is the realization that none of the particles that have been introduced so far required a curved or round shape to make their structure work.

This became clear already in the review of the supreme symmetry of the proposed pyramid squares numbered 2, 4, 5, 6 and 8.

The above therefore strongly suggested that the proposed pyramids might be built up by bricks that would have a rectangular shape, box like, but again then, for the sake of supreme symmetry, this brick would have to be a cube, and a very small cube at that.

Because the baselines of the pyramids that were devised here all had a directional arrow in them, I felt that these cubes, with a little directional arrow in each of them, could be arranged in a parallel linear configuration.

How could these cubes exist?

What we need to do is to go back to Fig. 1, where it was argued that a whole network of crisscrossing, alternating plus and minus signs were actually representing a perfect ZERO, NOTHINGNESS.

We can do that again here:

Figure 18 is an arbitrary large cube that is a sampling of a piece of the 'vacant' universe, which is built up of pairs of small cubes that face each other with <u>opposing</u> arrows, which means that such an opposing pair represents nothing at all, it's a zero. In order to have each little cube preserve its own individuality, it was necessary to have these small cubes not touch each other along the full length of their individual sides, but only touch each other at their corners.

The outcome of this was that all cubes were now separated on all of their six individual sides by empty cubic spaces.

The small cubes are aligned in rows in the large composite cube. Imagine now that each small cube that points its arrow now towards the reader of this page would have its arrow turned 180 degrees around, away from the reader, then we would see in Fig. 19 a bundle of aligned cubes which all point in the same direction, away from the reader.

This group of arrow cubes has now become a bundle of aligned rows which are pointing their arrows all in the same direction.

These active aligned forces in these pyramids represent the <u>INTRINSIC SPIN</u> of particles.

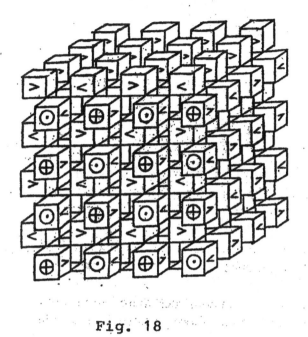

Fig. 18

Fig. 19

In this theory there are no strings or looped strings that can go through all sorts of contortions. This is not a string theory.

However, drawings 18 and 19 display a pair of cubes which can be considered to consist of bundles of stiff single strings, but all of these strings are straight, very straight, like long arrows.

There are no circles or curves here, every arrow in this system has either a parallel or a perpendicular relationship with all other arrows of the particle group it is in.

This theory is about the structure of particles which are alive and operate in a spatial matrix that completely controls their behavior.

Spacetime is something entirely different, it will be reviewed here later on.

Figure 20 indicates how a pyramid can be constructed with cubic bricks so that its pyramid slopes can have a 45° angle with the pyramids baseplate.

The heavy black line represents a 45 degree cut that can be made through a large cube that is full of small arrow cubes which are separated by small vacant cubes.

The resulting 45° slope is the only slope that this system can construct, there is no room for error, and it is the main key to nature's ability to reproduce those particles it creates in an entirely consistent manner.

There are no smooth sides on any pyramid slope in this theory, there are not flat slopes.

As to the arrows in these pyramids:

The proton and neutron structures get their particular character through the merging of two neighboring pyramid slopes at various locations of their double pyramid groups, and it is by means of these connecting pyramid slopes that stronger, modified forces are created at several locations, as well as electric charges, gravity, etc.

This will be shown further on.

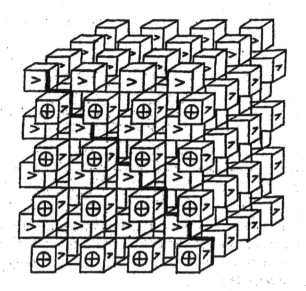

Fig. 20

At this point it became necessary to draw pyramids which consist of small cubes with individual arrows.

For the sake of symmetry and balance it seemed logical to assign an even number of cubes to each baseline. This even number would thus provide for the possibility to split a baseline in half, if possibly needed, so that this baseline's left side and right side would have an equal number of cubes each.

Fig. 21 shows how this vertical dividing line makes two halves. A horizontal cut like that also makes two equal halves.

With that arrangement as a basis, I attempted the remainder of this theory to fit in, by trying to find designs, properties and characteristics that could be argued to be viable in the basic concept of this theory in which the characteristics of the neutron, proton, electron and neutrino would be reflected.

It did not happen. After about a decade I had to change gears.

Fig. 22 shows an uneven number of small cubes in the pyramid's design in its baselines, which of course has the same number of cubes that one would find if a pyramid would be vertically cut right through its middle as is demonstrated in Fig. 23 with its cross-section ABCD.

This is the small cube arrangement that made it all work!

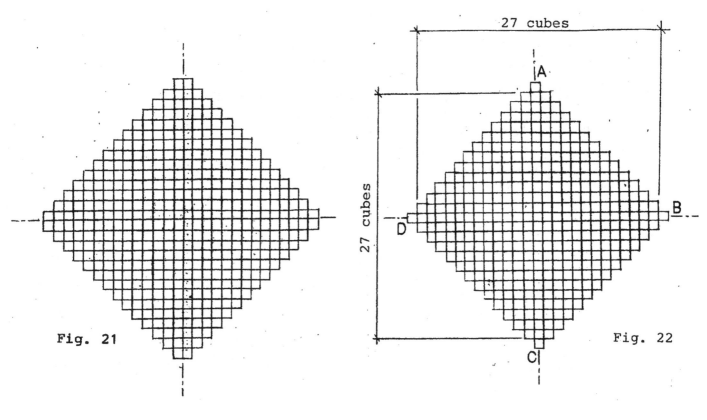

Fig. 21

27 cubes

27 cubes

A

B

D

C

Fig. 22

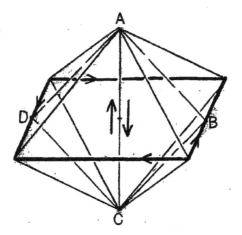

Fig. 23

The assembled five pyramid squares of Fig. 24 show, as seen from above, their arrow patterns of the small cubes.

Note that pyramid squares 2 and 8 show their CCW rotation, and that pyramid squares 4 and 6 show their CW rotation.

Central square 5 has conflicting arrows on its perimeter, its baselines have diverging directions in its NW and SE corners, and they have converging directions on its NE and SW corners, which feature will later be shown to create nuclear mass.

These arrows in this drawing number only 7 in a baseline. This number 7 was only a guess. Eventually it turned out that it happened to be correct.

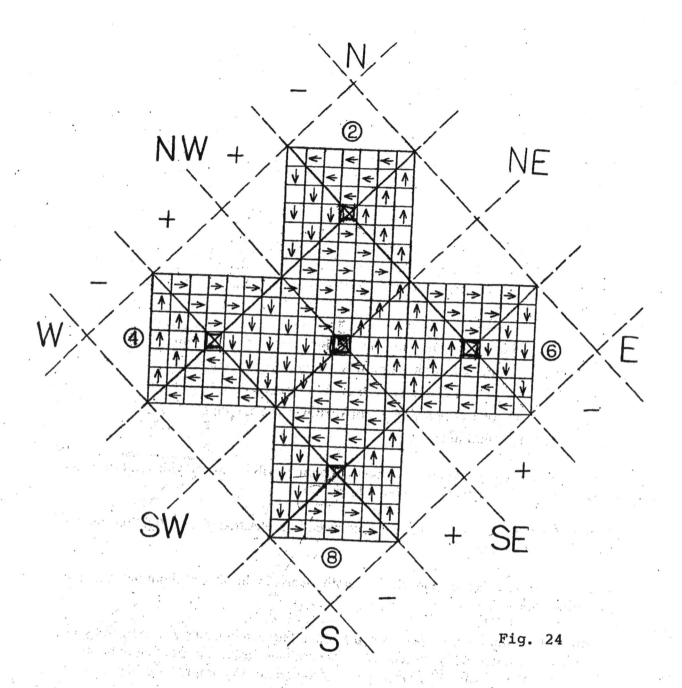

Fig. 24

The bottom to bottom pyramid assembly that has been displayed here has a characteristic that was built in from the very beginning when I had reasoned that the slopes of the pyramids should have a 45° slope with their pyramid's base so that these pyramids could be packaged nicely and tightly together if the need might arise to do that.

For more than a dozen years I was not aware of the hidden implications of that, which is: that the length of the common centerpost that connects the two tops of the two pyramids is not only equal to the length of each of these pyramid's baselines, see Fig. 25, but that the isosceles triangle of the pyramid slope that faces the reader in Fig. 26a is actually congruent with Fig. 26b's isosceles triangle that has the pyramid's centerline as its baseline.

> **This means that the combined centerposts of the two bottom to bottom pyramids can and will function as a baseline of a pair of bottom to bottom pyramids that will be a true <u>inside out</u> offspring of the original pyramid pair.**

From here on forward, all double pyramid's baselines will be drawn with solid lines, their centerposts will be drawn with dash/dot lines.

The above described congruence of the indicated isosceles triangles in Fig. 26a and 26b has monumental implications.

It provides the foundation upon which everything else in this theory about the structures of particles in Particle Physics is based.

An original pyramid in Fig. 27a is shown again in Fig. 27b in which it demonstrates its capability to lend its centerpost and one of its baselines for the construction of a true inside out double pyramid that has now one of the original pyramids' baselines s its centerpost.

The ultimate achievement is the creation in Fig. 27c of four inside out pyramid pairs around the single pyramid pair of Fig. 27a.

There are examples of this inside out feature in biology: Cows have their skeleton on the inside and their meat on the outside, a lobster's skeleton is on the outside and its meat is on the inside.

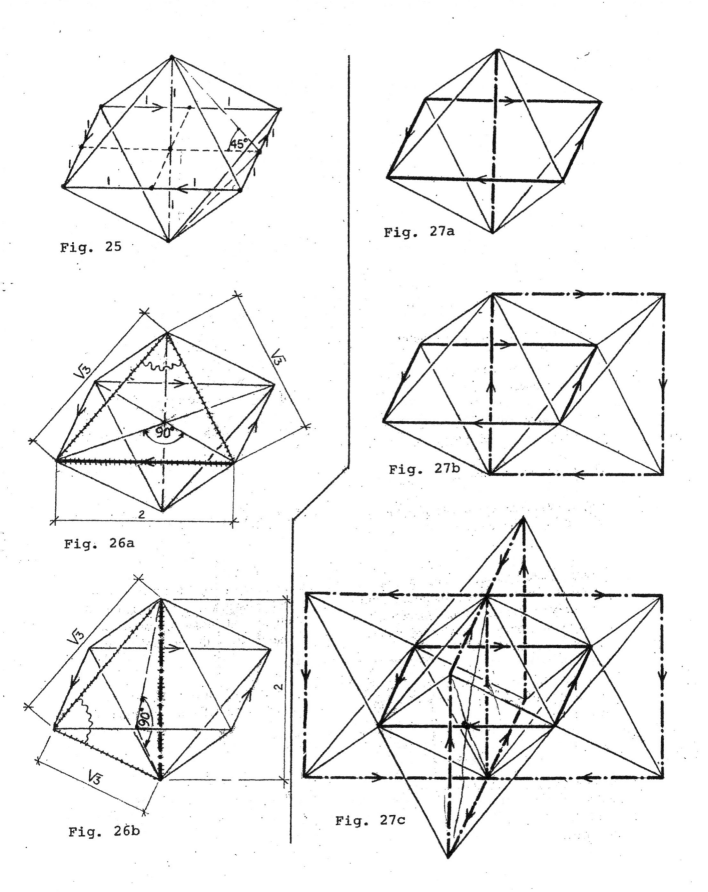

Fig. 25

Fig. 26a

Fig. 26b

Fig. 27a

Fig. 27b

Fig. 27c

33

The cross-shaped assembly of the five pyramid pairs that we have dealt with so far are actually assembled tightly in a stable and balanced package that has its center of geometric gravity exactly midway between any identical but diametrically opposite components.

Fig. 28 shows how the pyramids of square 2 and those of square 4 are both folding upward till pyramid slope 2W merges with pyramid slope 4N so that now the joined baselines of those slopes are now pointing straight down on square 5's NW corner in Fig. 29. Fig. 28a indicates how the folding (imaginary) hinges work.

For the sake of perfect balance, the pyramids of square 6 and square 8 have to fold downward, till slope 6SO merges with slope 8EO. Their baselines are now straight under the SE corner of square 5, and pointing upwards.

Fig. 29 now shows the real structure of the neutron, as proposed in this theory.

The baselines of this assembly have now started to form cubes, and all five centerposts are now connected in a way that also indicates that they might become a part of a cube, which is indeed what will happen.

Square 5 in the center is now tightly surrounded by structures which later will be shown to be deeply involved in the information of nuclear mass.

The quarks will be identified.

The arrows in the centerposts of squares 2, 4, 6 and 8 all show their respective CW relationship with their surrounding baselines.

Central square 5 however shows two opposing arrows in its centerpost, due to there being a conflict of arrow flows in the four baselines: If the centerpost arrow goes upward, then it has only a CW rotation relationship with square 5's left and right baselines; when the centerpost's arrow goes downward, it only has a CW relationship with square 5's front and rear baselines.

NOTE:

Eventually the outcome of this conflict was that this centerpost was identified as being devoid of arrows, but that it was found to be a 2-way street where the vacant state of the center post allows an arrow flow in either direction, as shown in fig. 29.

When reading this, always keep in mind that when the arrow in a baseline is discussed that we are always talking about that <u>one quarter section</u> of a bottom to bottom pyramid pair that has <u>all of its rows of arrows</u> going in the same direction as that of the baseline on their perimeter.

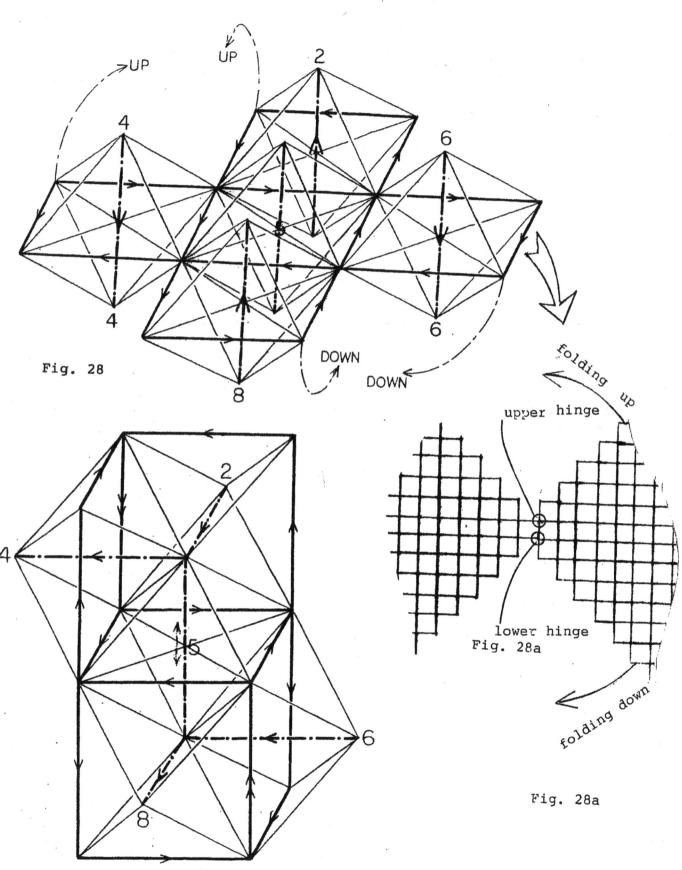

UP

UP

2

4

6

Fig. 28

8

4

DOWN

DOWN

6

folding up

upper hinge

lower hinge
Fig. 28a

Fig. 28a

folding down

2

4

5

6

8

Fig. 29

35

Figures 30 and 31 are typical examples of how these patterns can extend themselves into squares which are all united in cubes.

In Fig. 30 all of its solid lines are baselines of pyramids, and all dot-dash lines are the original centerposts that are all getting interconnected with each other, forming also cubes.

In Figure 31 a cube is subdivided into six single pyramids, which can be done with every cube.

The important thing is that the baseline cubic system and the centerpost cubic system are completely interwoven so that the offset between these two cubic systems is exactly equal to half the length of a cube's side. The result of this is that the corner of a cube of one system is located in the exact center of the other cubic system.

In this theory each particle such as the neutron, proton, electron, neutrino, etc. is subject to this system of two interwoven cubic lattices which have an unchangeable geometric position between them.

Particles are made of whole or partial pyramids, or combinations thereof. Pyramids are made by connecting a corner of a baseline cube with a corner of a centerpost cube, in any direction it chooses to do so.

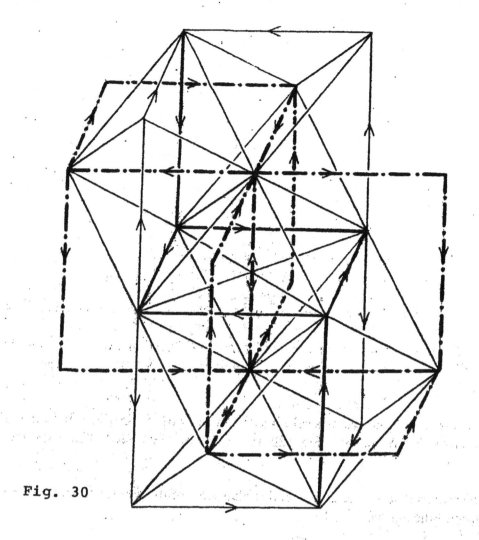

Fig. 30

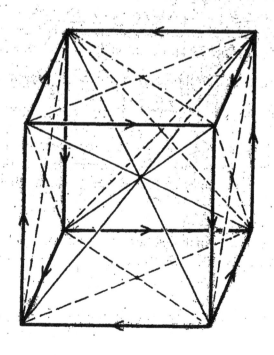

Fig. 31

Fig. 28 and 29 have shown how the five bottom to bottom pyramid pairs had folded themselves against each other so that they would become a more compact package.

When being on its own, a free neutron is known to split itself after about fifteen minutes into a proton, electron and neutrino.

The neutron's break up of Fig. 15 is shown here again, Fig. 32a. There is a good reason why the break up occurs at those two particular locations, shown in Fig. 32c.

Fig. 32b indicates in which directions the arrows are pointing.

Fig. 19 shows how arrows which all point in the same direction are being assembled side by side and how they are stacked vertically. In the neutron's matrix it is done exactly the same way.

Fig. 15a and 29 have shown that the NW-SE diagonal is the commanding direction along which all of the arrow cubes are concentrated. After the creation of the neutron, all of the pyramid slopes of sections 2, 4, 5 and 8 had connected slope on to slope, making a tightly packed assembly, EXCEPT those four V-shaped half double pyramids which in Fig. 29's neutron structure stick out by themselves into space, unsupported, weak and fragile, and after 15 minutes they drop away as in Fig. 32c, thereby changing the neutron into a proton, while the four V-shapes make an electron, which immediately thereafter will lose nearly all of its components, keeping only those few building blocks which are joined tightly together in the electron's center. More about this in "The Electron" chapter, Fig. 40-41.

As this theory develops further, we will discover that this proton and electron themselves also have rather feeble components, which, after their drop off will assemble into a neutrino.

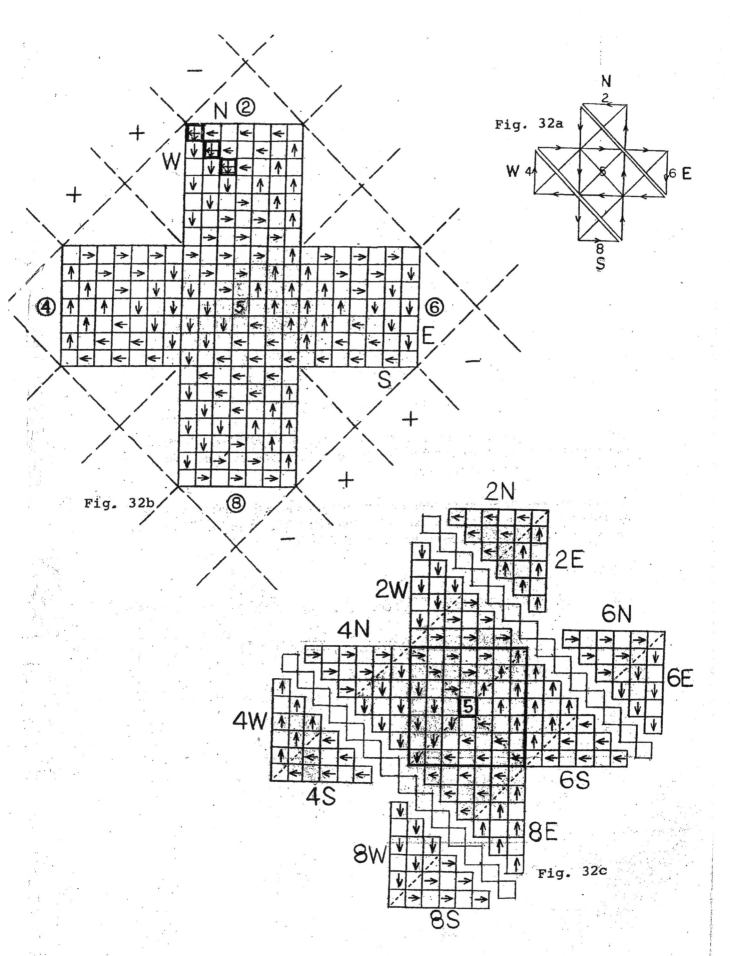

Fig. 32a

Fig. 32b

Fig. 32c

39

Prior to its birth the proton occupies the central portion of the neutron, which is more apparent and easier to see in the neutron's break up in Fig. 33a.

Sections 2, 4, 6 and 8 all lose their outside halves which will later combine to construct the electron.

When the neutron is in its compact, folded configuration, it has not in any way affected the weak diagonal links' tendency to break apart, because no restriction has been placed on the neutron's capability to break apart at its designated diagonals in the NW/SE direction.

Figure 33b represents the compact, single neutron, its to be ejected components that will later construct the electron are drawn in dashed lines.

The neutron's very stable remainder is the proton, Fig. 33c.

What we see here now in the proton as well as in the neutron that the once 'vertical' centerposts of sections 2, 4, 6 and 8 are now positioned 'horizontally,' connecting to the centerpost of centrally located section 5.

These internal centerpost connections are of major importance, this makes the whole system work, without these connections there would be no proton, period. There would be no neutron either. The following pages will explain that.

Fig. 33b and Fig. 33c demonstrate the SUPERSYMMETRY of the neutron and proton.

Supersymmetry in physics is like Siamese twins who are connected at the top of their head and are looking in opposite directions.

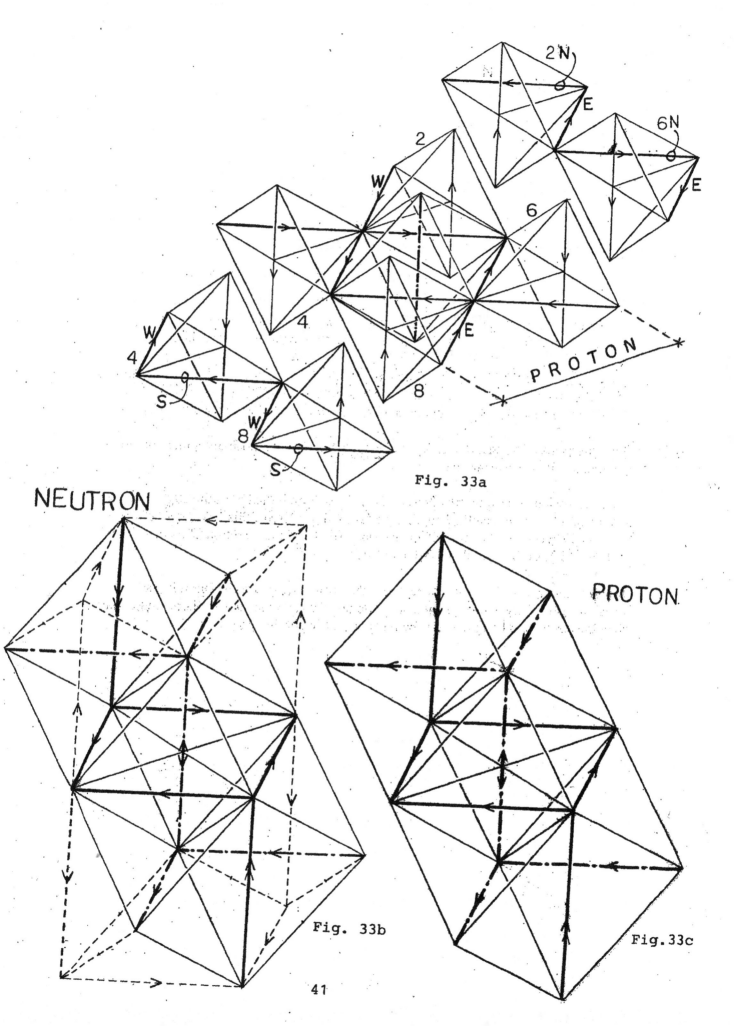

Fig. 33a

NEUTRON

Fig. 33b

PROTON.

Fig. 33c

41

Figure 34a displays the pyramid slope identification system that was first introduced in Fig. 3, 6.

When this proton is flipped end over end then we see its other side in Fig. 34b.

The solid line baselines and the dash-dot centerposts are in both drawings in exactly the same place, everything is congruent.

All baselines have a directional arrow which indicates the direction of all force arrows in that particular bottom to bottom pair of quarter pyramid slopes.

Section 2, 4, 6 and 8 all have a baseline arrow system that indicates a CW relationship with this distinction that section 2 and 8 have a rotational direction in the flat matrix of Fig. 32b that is opposite to the rotational direction of the baselines in section 4 and 6.

The centerposts of all sections 2, 4, 6 and 8 all accommodate those particular conditions by being clockwise everywhere.

In the discussion regarding Fig. 29 as to the CW rotation indicator in the dash/dot lined centerpost, it must be recalled here that the original premise of this CW rotation was its reference between the direction of an electric current and the direction of its magnetic field around it, as first stipulated in the explanation of Fig. 7.

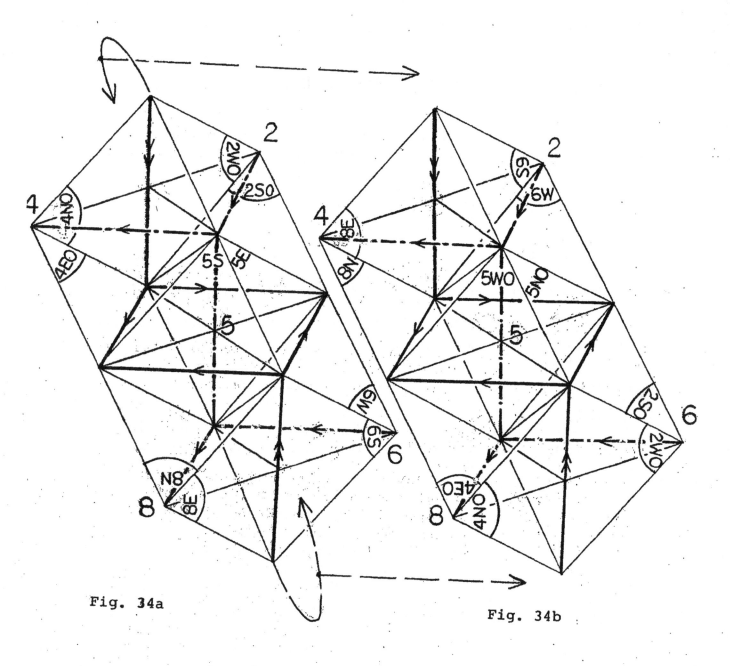

Fig. 34a

Fig. 34b

The structures of the neutron have already suggested that the geometric patterns of their structures might be suitable to be extended into its surrounding space. Fig. 30 and 31 both seem to indicate that the patterns that have been used so far are a small sample of our surroundings.

This is not to say that surrounding space is permanently fixed in an unmovable pattern.

It is the other way around:

Space, just by itself, is just space, unstructured, but capable to transfigure itself on less than a moment's notice into a pattern that is needed by the particle or wave that enters its realm.

This means for instance that a lightwave, a photon, is actually not traveling itself through space, instead, what happens is that a photon hands over its energy to a particular space pattern that is exactly suited to pass on this energy to the next space pattern, and so on and on.

A sample of this propagation will be shown later on.

When particles in space interact with one another, such as an electron orbiting a proton, or two protons and two neutrons connecting to make helium, then those assembly patterns will fit exactly in their own prescribed internal prefabricated structure that is made according to the unbending rules of the spatial matrix of Fig. 35.

****

THE BIGGEST ASSET OF THIS THEORY IS THE UNEXPECTED REPEATABILITY OF THE INTERLOCKED SPATIAL ARROW PATTERNS OF THE NEUTRON, PROTON, ELECTRON, NEUTRINO, REACHING ALL THE WAY INTO INFINITY, THEREBY CONTROLLING WITHOUT FAIL ALL INTERACTIONS BETWEEN PARTICLES.

****

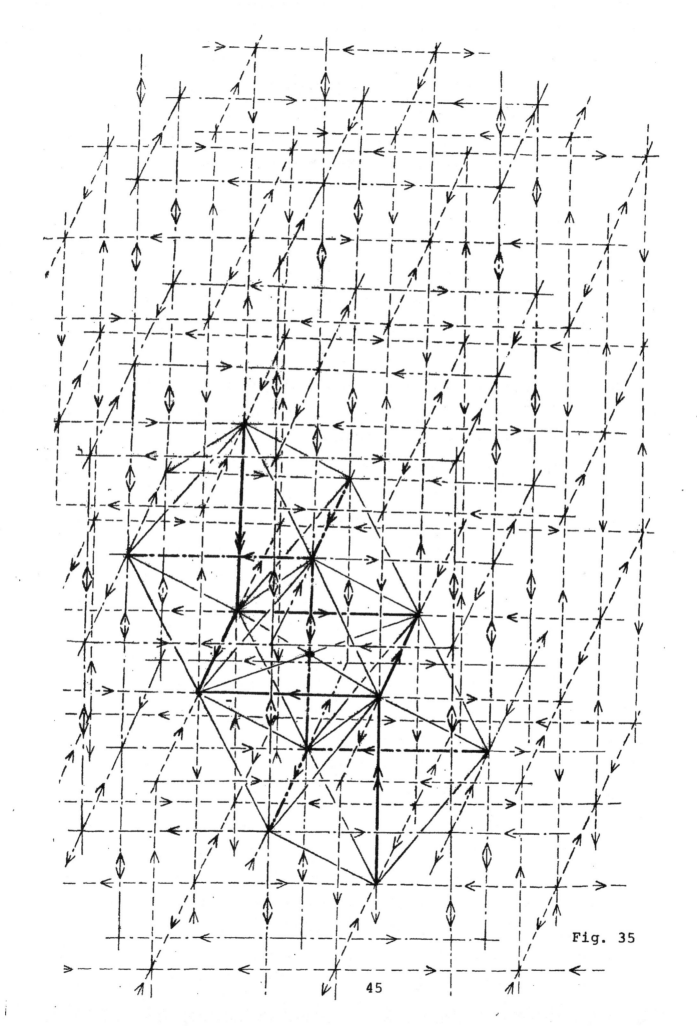

Fig. 35

When, for study purposes, we cut the proton of Fig. 36 into equal halves, then it becomes clear that each half, Fig. 37a and Fig. 37b, consist of two standard, interlocked XYZ coordinate systems.

The origin of these solid-lined pyramid baseline XYZ coordinate systems are located in the exact center of the cube that is made by the dash-dot lined pyramid centerposts.

. . . In reverse:

The origin of these dash-dot pyramid centerpost INSIDE OUT coordinate systems are located in the exact center of the cube that is made by the solid-lined pyramid baselines.

In a few words:

These two interlocked coordinated systems are each other's inside out. They are at the same time cooperating and competing with each other.

This back and forth interplay makes these systems work.

It also makes this theory work.

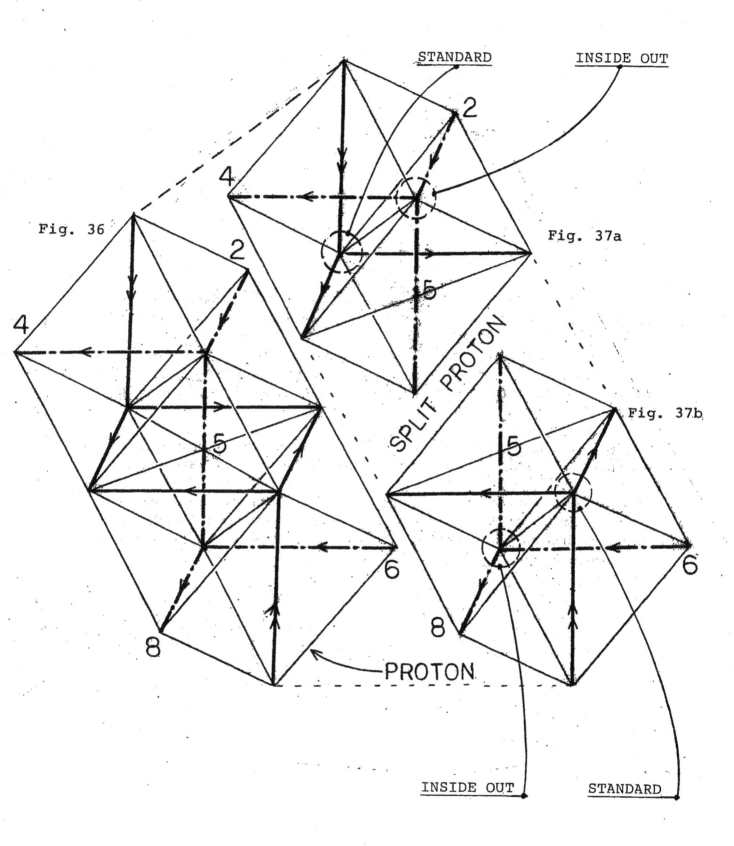

STANDARD   INSIDE OUT

2

Fig. 36

Fig. 37a

4

2

4

5

SPLIT PROTON

Fig. 37b

5

5

6

6

8

8

PROTON

INSIDE OUT   STANDARD

The positive charge of the proton and the negative charge of the electron are tightly connected to their respective home sites. All along in this analysis it was a real problem not being able at its very beginning to identify their particular pointlike location as well as the localized conditions that would render a location either positive or negative.

The initial proposal in Fig. 15 where plus and minus signs were placed in the respective proton and electron sections turned out to be exactly correct, but even so it was no indication if an electric charge was located in just one baseline, or in a corner where two baselines connected, or just on the slope somewhere on a pyramid, or whatever.

With Fig. 35 with a proton within its matrix as a reference, it was found that a positive charge is produced where at an intersection one baseline flows in and four baselines flow out. With a negative charge it is exactly the opposite: one baseline flows out and four baselines flow in. Any structural corner where less than three baselines meet does not qualify.

Fig. 38 shows that the proton has two positive sites, due to the fact that it is a balanced structure that has two equal candidates for being positive at opposite sides. These two charges are in parallel, not in series.

Fig. 39 is the neutron in its actual packaged configuration. When the neutron breaks up into a proton and an electron, then immediately thereafter the proton and electron will both eject two small components each, which will then combine to make the neutrino. More about that later.

The above paragraph serves as a reminder that it is just highly improbable that the neutron is completely devoid of positive or negative electric charges, internally.

These charges are just not suddenly fabricated when the neutron disintegrates, they were always there in the first place, and their presence within the neutron was such that their internal distribution within the neutron provided for a perfect mutual neutralization as to their inherent electric charges.

It will be explained later that the two extremities of the proton shown with wiggly lines in Fig. 38 will break away from the proton, and what that is.

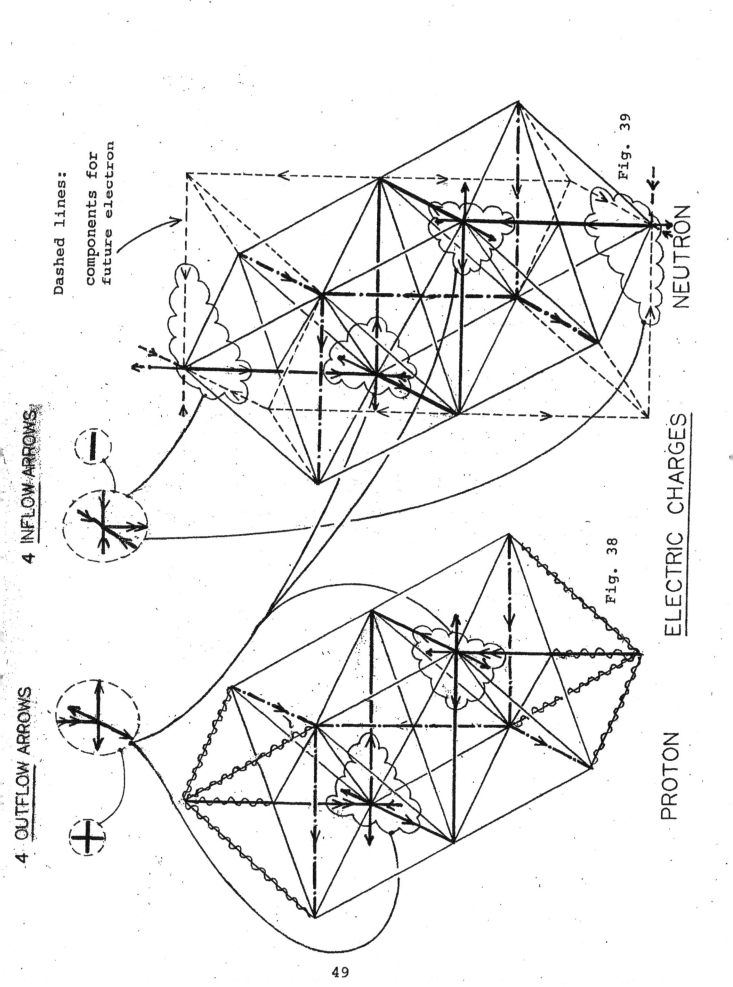

4 OUTFLOW ARROWS

4 INFLOW-ARROWS

Dashed lines:

components for
future electron

Fig. 38

Fig. 39

PROTON

NEUTRON

ELECTRIC CHARGES

49

Fig. 24 displays what the neutron looks like in its as yet unpackaged form, being five double pyramids in an arrangement in which a single pyramid pair is surrounded and locked in by four other pyramid pairs. See Fig. 22.

When the neutron's split up of Fig. 33a happens, it splits in a manner as drawn in Fig. 40.

The four V-shaped split-offs that leave the neutron's remains are shown in a planview-like flat shape, but thereafter, a little farther out they are shown in 3-D, with their pyramid slopes identified.

Following the drawing's arrows, at two locations a larger particle will be formed by two smaller ones, when those pyramid slopes which have been shaded for the viewer's convenience, are merging.

Each of these two entities now represent one half electron, and they are identical, as if they were clones, as described below.

What we see now is the assembly of four quarter electrons into two half electrons, each of which now has two single inward flowing arrow baselines and one doubled up pair of outward flowing arrow baseline.

These two half electrons will then find each other and merge, on the next page.

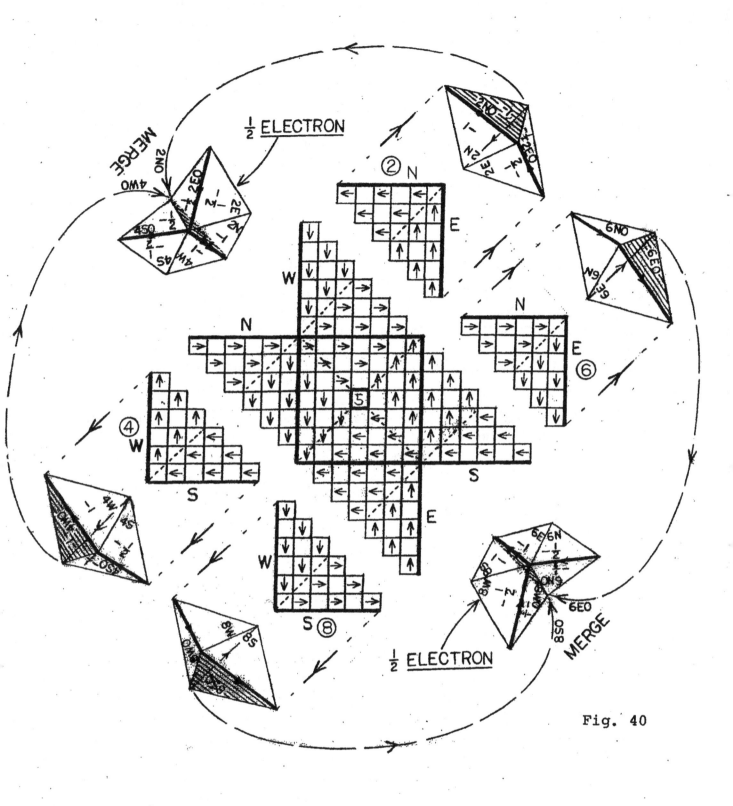

THE NEUTRON'S BREAK-UP.

Each of the neutron's four discards consist of two bottom to bottom half pyramids, which will merge into two identical pairs.

Each one of these pairs is a half electron.

The two half electrons of Fig. 40 will merge here in Fig. 41a, thus producing the cross-like structure of Fig. 41b. Each half electron contributes its in-moving arrow baselines and its outflowing arrows to make quadruple packaged outflowing baselines which feed four single inflow baselines. We now have an electron, and it is negative.

This Fig. 41b structure of the electron is nearly half as big as one of the half protons of Fig. 37a,b. That cannot be of course, the electron is supposed to be no more than a "point like" particle. What is going on? I struggled with that for many years, unwilling to abandon this electron's logical configuration.

In September 2006 the answer just popped up: The electron that I had 'found' was a heavily accelerated electron, which might have a nuclear mass as big as that of a half a proton. But the electron that was ejected by the neutron's break-up should then be an electron with a minimal structure, as small as possible, with just a nuclear mass of one, being 1836 times smaller than that of a proton. For more on this, go to Fig. 68a,b.

Reducing Fig. 41a's two large half electrons to their minimal configurations produced Fig. 41c, d, e, f, which consists now.

At those negative charge 'spots' that can be found at the very bottom of Fig. 39's upper and lowest locations on the neutron.

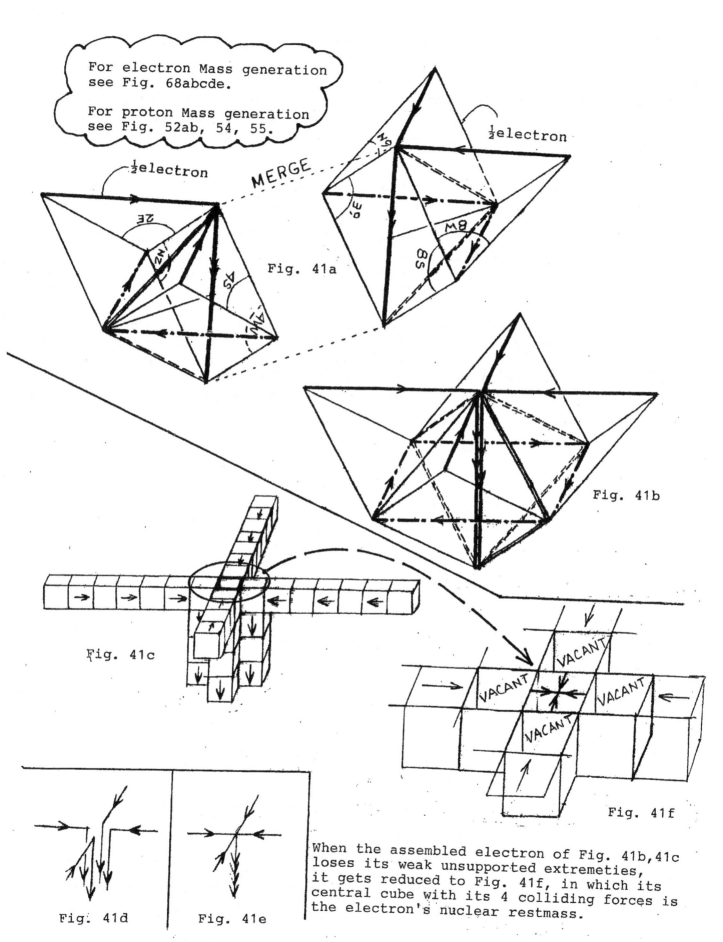

For electron Mass generation see Fig. 68abcde.

For proton Mass generation see Fig. 52ab, 54, 55.

½electron

MERGE

½electron

Fig. 41a

Fig. 41b

Fig. 41c

Fig. 41d

Fig. 41e

Fig. 41f

VACANT
VACANT
VACANT
VACANT
VACANT

When the assembled electron of Fig. 41b,41c loses its weak unsupported extremeties, it gets reduced to Fig. 41f, in which its central cube with its 4 colliding forces is the electron's nuclear restmass.

A comparison can be made now between the proton and the electron. Fig. 42a shows the proton and how it has been dissected, showing its identical separated end positrons which carry its positive charge. Next to these two proton ends are a half electron, each with its own negative charges.

When we now compare each separated proton end with the half electron, we see then that they are identical, except all of their arrows are reversed.

When we now assemble the split half protons back to back, then we get the positron of Fig. 42c. This type of assembly method is an exact copy of how in Fig. 41a, 41b the electron was assembled. The positron's rest mass is again produced by the crowded interference of four force arrows.

NOTE: In the very beginning of this manuscript it was stated that the mutual annihilation of an electron and a positron was the main reason for starting this theory from an absolute zero. The great benefit of that kind of a start was that I was not bound at all by conventional, existing theories. This freedom allowed me to use all of the information that was gathered by others in untried ways.

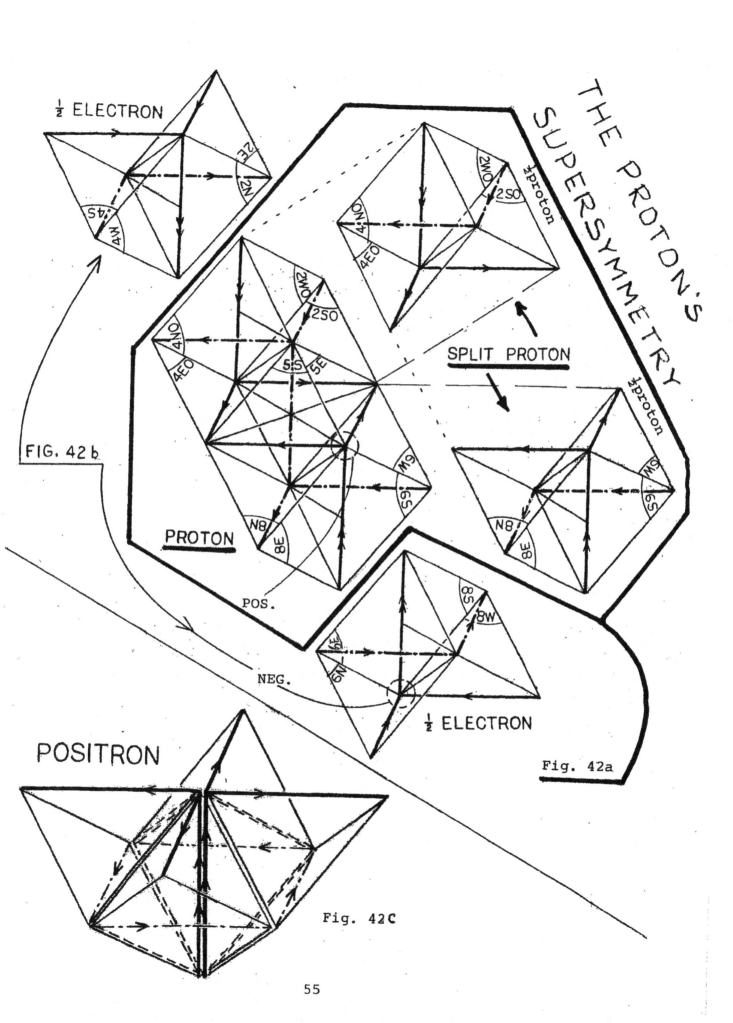

½ ELECTRON

THE PROTON'S
SUPERSYMMETRY

½ proton

SPLIT PROTON

FIG. 42 b

½ proton

PROTON

POS.

NEG.

½ ELECTRON

Fig. 42a

POSITRON

Fig. 42C

55

This most elusive of all particles was very difficult to find in this theory, but when it finally was found it also cleared up some other issues that were still outstanding with the proton and electron. As usual, as it always has been in this analysis, it had to do with structure, stability and balance. Nature does practical things, with great precision and repeatability, how come that the magnitude of every proton's nuclear mass is always exactly the same?

The neutrino owes its birth to the initially structural instability of the proton and the electron, which both have exactly the same flaw. The flat diamond shaped extremity which was reviewed in the electron's structure as well as in the proton's analysis is actually more or less a point that sticks out in space. It has no buttress of any kind to give it support, and it also is not symmetrical in its front/back configuration that would give it some equilibrium in its immediate vicinity. It seems therefore very plausible that for those reasons this pointed projection breaks away from the proton as well as from the half electrons, Fig. 43a and 43b. The result of this is that the half electrons and the two proton ends now have a somewhat 'rounder' structure, they have become tighter knit and therefore more stable.

(NOTE: What led me to this 'break-away' scenario as being a real event that would involve every proton, was the intermittent but continuous search for the location and shape of the quarks. With the above described piece of proton broken off, it would solve the quark problem, it would solve the neutrino problem, and also the gravity and temperature problems.) It was probably for similar reasons that the neutron lost its 'pointy' projections, being those four half bottom to bottom double pyramids that it released.

In Fig. 43a, subject half pyramid pieces that break away from the proton are marked with PN (=proton/neutrino). In Fig. 43b, subject half pyramid pieces that break away from the two electrons are marked with EN (electron/neutrino). These four particles will now assemble and become one on the next pages.

We must keep in mind, in this case in particular, that the baselines are <u>single row</u> cube columns which actually contain arrow cubes, alternating with vacant cubes, and then of course the pyramids are all filled with rows of alternating cubes, but the centerposts of pyramids are always empty tunnels, even after they have been folded over and take baseline-like positions in an inside out pyramid formation.

This will become apparent in the neutrino's formation.

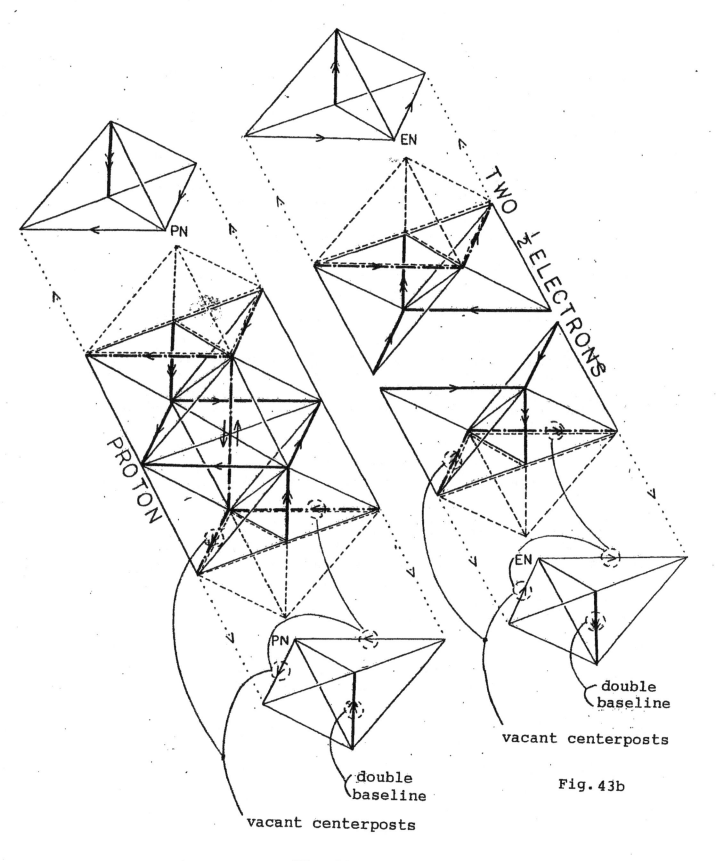

PN

EN

TWO ½ ELECTRONS

PROTON

PN

double
baseline

vacant centerposts

Fig. 43a

EN

double
baseline

vacant centerposts

Fig. 43b

57

In Fig. 44 the original five double pyramid squares which made up the neutron are recalled here in order to shed light on what happens to their baselines as they fold up or down.

Baselines 2W and 4N will fold up, but their baseline arrows will then be pointing down. Baselines 6S and 8E are both folding down, but their arrows will be pointing up.

Taking now the two proton cut offs of Fig. 43a, shown here assembled as Fig. 45a, and the two electron cut offs of Fig. 43b, shown here assembled as Fig. 45b, they can then be assembled in Fig. 45c with the stipulation that of course their centrally located arrows of their half baselines all point in the same direction.

The result of this assembly is a double, bottom to bottom pyramid with vacant baselines.

IT IS A COMPLETE <u>INSIDE OUT</u> PYRAMID PAIR.

There are now <u>four</u> real baselines acting as a single centerpost in the center of these bottom to bottom pyramids, and, combined with further analysis of behavioral details of an electric current, yet to come, where an assembled centerpost also has four half baselines, a conclusion came forward that indicated that such a combination of those four packed together and centrally located baselines empowers these particles with its capability to pierce its way at great velocities straight through other particles.

Except for the missing four baselines, the complete neutrino model of Fig. 45c resembles the shape of the bottom to bottom pyramid pairs of sections 2, 4, 5, 6 and 8, except that it is an inside out pyramid pair.

As such, the neutrino as drawn here has quite a large circumference, and having mentioned the subject of particle stability in these pages several times, it seems logical to suggest now that shortly after its birth the fast traveling neutrino will quickly be stripped of its excess baggage as it traverses all sorts of other particles that are in its way.

That excess baggage are all of those rows of parallel arrows which don't have the benefit of the reinforcement of a nuclear mass structure.

Being slimmed own like that, these skinny remains will have little or no trouble to penetrate its obstacles like a spear, Fig. 45d.

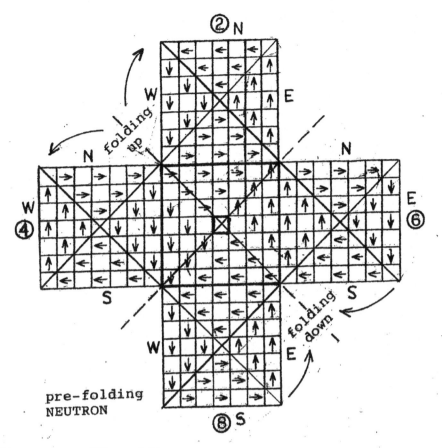

pre-folding
NEUTRON

Fig. 44a

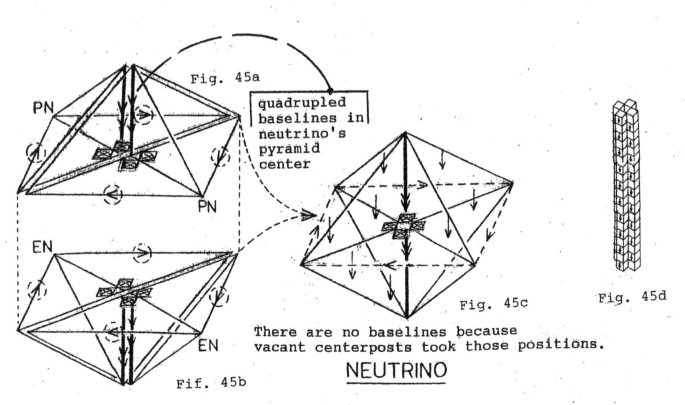

Fig. 45a

quadrupled
baselines in
neutrino's
pyramid
center

Fig. 45c

Fig. 45d

There are no baselines because
vacant centerposts took those positions.

NEUTRINO

Fif. 45b

59

The subject of gravity is as old as the world. Philosophers and scientists have usually viewed gravity as the original source of everything else in the world. Even so, gravity still seems to be the least understood phenomenon, in particular because it has not been found to fit in the STANDARD MODEL OF PHYSICS. In the early phases of the development of this theory, in Fig. 15, after only a few days of evaluating the possible significance of the criss-crossing lines of Fig. 3, it seemed to be a reasonable guess, for starters, to suspect that the pull of gravity was located in quarter pyramids 2W/2WO, 4N/4NO, 6S/6SO and 8E/8EO, with their inward arrows.

It has been a tortuous path, full of efforts going nowhere, with small advances that seemed to point in the right direction, then long periods of inactivity on this particular subject, then a switch to other quarter pyramids that somehow would seem to hold more promise, exploring this new direction, hitting a brick wall again, working on related issues, leave it alone for awhile, and so on. In the end, there came an answer that works. Here it is:

As a source of building materials for the neutrino, it was shown in Fig. 43a that the proton's projecting portions of its body were cut away, being the upper halves of quarter pyramids 2W/2WO and 4N/4NO, as well as similar halves of quarter pyramids 6S/6SO and 8E/8EO, shown here shaded in plan view for the whole proton in Fig. 46a.

As the formation of this theory progressed, it became apparent that the hard, tough portion of the proton consists of its central section 5 that was supposed to hold its nuclear mass, as well as those quarter pyramids which have boxed in that nuclear mass from the outside inward, being 2S, 4E, 6WO and 8NO.

The remaining halves of the above-mentioned quarter pyramids 2W/WO, 4N/4NO, 6S/6SO and 8E/8EO are marked in Fig. 46a with an 'R,' they have lost a structurally supporting partner, but it opens the door for the proton to perform all of those functions that it is known for, and which the neutron never had. To keep the explanation simple, see Fig. 46b, with section 2 only. Section 2's half of tetrahedron 2W/2WO is here identified as 2W-N, meaning it is the North half of said tetrahedron, which it had lost to the neutrino in Fig. 43a and 45a, and when we take it, in our mind, at point P, then we can rotate it around the noted pivot axis, and deposit it at 2E-S, with which it is congruent, in which it fits exactly, even the baseline arrows fall correctly in place. The only (important) difference is that gravity's inward pointing arrow has now become the outward arrow of the weak force, in Fig. 46c.

Fig. 46c is a simplified drawing of section 2, as it relocated quarter pyramid 2W-N (with its wiggly lines) into vacant quarter pyramid 2E-S, with its shaded area.

How can that be, how can nature justify this?

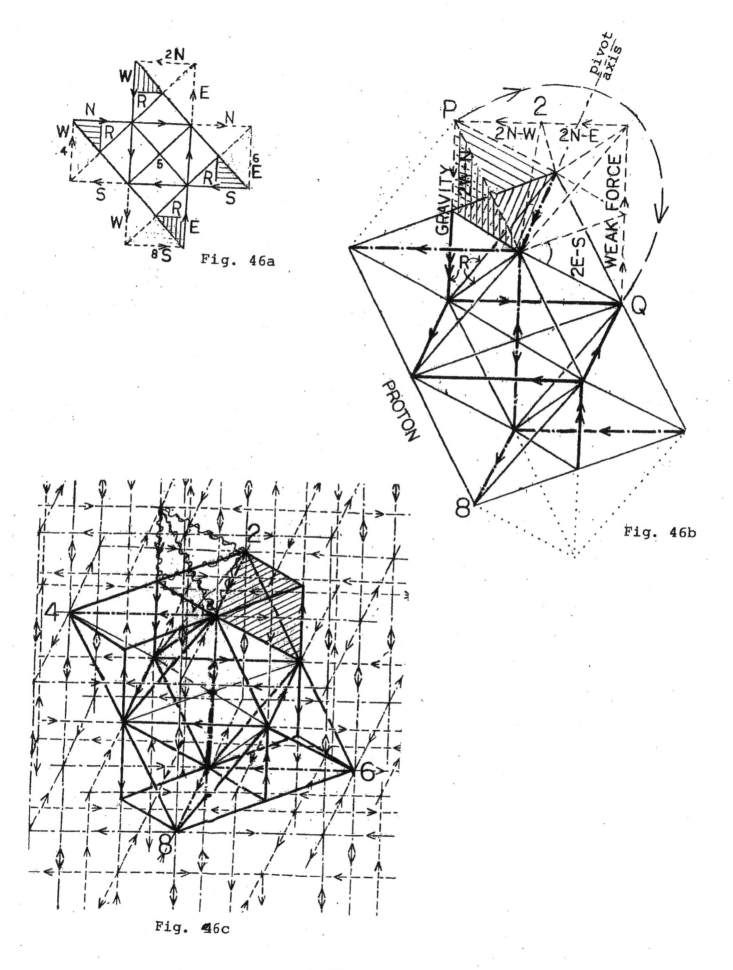

Fig. 46a

Fig. 46b

Fig. 46c

For the purpose of evaluation, we can take an algebraic/geometric approach to our understanding of this modification of the proton, again only shown here for section 2. This review is split in two parts, Fig. 47.

$\frac{1}{4}$ pyramid 2W-N (GRAVITY)	$\frac{1}{4}$ pyramid 2E-S (WEAK FORCE)
inward arrow … is negative.	outward arrow … is positive.
This is about THE LOSS OF A NEGATIVE ITEM	This is about THE GAIN OF A POSITIVE ITEM.
Result: A NEGATIVE LOSS IS A GAIN.	Result: A POSITIVE GAIN IS A GAIN.

THESE ARE IDENTICAL STATEMENTS!!

In other words, the proton did not lose anything, it just remodeled. What we are seeing here is that nature is structurally operating on both sides of the zero line. It does so in its inside out relationship between the proton and its electron, it does so within the confines of the neutron, it does so everywhere, always, as will be shown as this theory develops further.

At this point it is a fair proposition to state that the density of the proton's structure, when one goes from the inside to outside, decreases, starting at the center of Fig. 48. There is no doubt that the proton's nucleus, within section 5, is the toughest, the hardest.

Going from there outward, then we encounter the next layer, which consists of tetrahedron 2S/2SO of section 2, tetrahedron 4E/4EO of section 4, tetrahedron 6W/6WO of section 6, and tetrahedron 8N/8NO of section 8. Theses components will later be shown to shelter the proton's quarks, which can then also be precisely identified, but they basically act as the mould, a softer but tough and resilient mould which on its exterior supports the third and outer layer of the proton.

These outer layers are to be found in those portions of the proton's sections 2, 4, 6 and 8 which do not have contact with section 5 at all, which are 2W/2WO, 4N/4NO, 6S/6SO and 8E/8EO. Those are exactly the components of the proton which are soft, and which can be manipulated, such as what happened when they were cut off to provide building materials for the neutrino.

There is yet another participant that belongs to the proton, which is the ever present matrix by which it is surrounded, which is like the air around us that we need to breath in order to be able to do the things that we do.

It is this matrix which serves as a gossamer extension for the exterior soft tissue of the proton.

In the end, these soft tissues are the undetectable constituents of space which manage to give us soft signals such as daylight, sound, hot and cold, gravity. The weak force might be called anti-gravity, and its interplay with gravity may determine whether an element is a solid, a liquid or a gas.

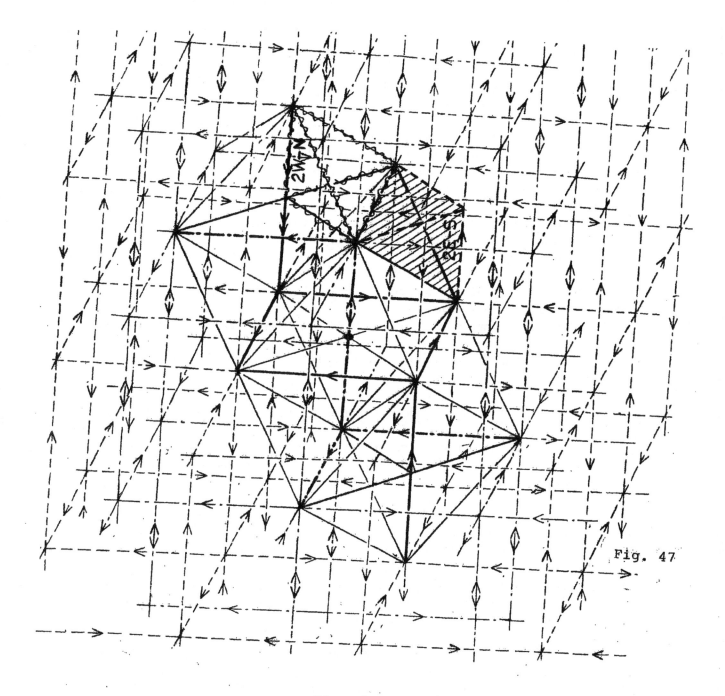

Fig. 47

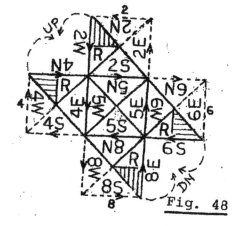

Fig. 48

The following explanations are intended to provide a comprehensive overview of a proton's capabilities and operating principles.

The explanation goes as follows, shown for one end of the proton only: In Fig. 43a the proton had lost its pointy extension, being a 'soft' portion of the proton. Thereafter the matrix fills it in with its desire to rebuild that lost portion. This rebuilding is somewhat accomplished by gathering yet free floating hot or cold energized single, small pyramid cubes from surrounding space which conform to the hot or cold condition of the proton at that moment. (This is the kind of reconstruction method that a split double helix is doing). For a hot proton this soft reconstruction attempts to fill in its part which it had lost to the neutrino.

In Fig. 49a, the hot proton is shown complete and fully stretched out, its double arrow gravity baseline at the proton's end is fully enclosed by the proton. Nothing remains here of those external matrix gravity components, the hot proton <u>swallowed</u> it up, it is gone, and because of that the proton lost its gravitational pull, it became a gas. This applies generally for a proton that is a part of a larger composite element, with neutrons, etc. Protons by themselves are at room temperature a gas already, most likely because their perfect structure is not compromised by attached particles.

In situations where the gravity's component is at least partly present in its attachment to the proton, then the proton still exerts a gravitational pull that is somewhat proportional to the length of this active, exposed portion of the matrix's gravity baseline. Fig. 49b shows the maximum gravity that we and the proton feel, which is the gravity that is NOT enclosed inside the proton's confines, but is the gravity of the matrix which at that particular location is attached to the proton.

Fig. 49b's embroidered diamond shaped scars, where the quarter electrons components used to be in section 2 and 4, are jointed at their central common border, in the hot proton they are large and diamond shaped, in the cold proton they have become smaller squares. In the hot proton the weak force is entirely outside the proton's enclosure, all of it is in a part of the matrix that is attached to the proton, and therefore very active, repelling its neighbors, while the gravity in Fig. 49a cannot be felt, it is locked up, making the proton a gas.

The hot and cold drawings of the proton in Fig. 49a and Fig. 49b are extremes, but there is a middle road, which can be anywhere between the hot and cold limits. See Fig. 49c.
The two connected embroidered diamond shapes resemble a pair of (very slowly) flapping wings of a bird. The dash/dot centerposts act as hinges for the diamond shapes: when their joined center pivots go up, their unattached outsides go down, and vice versa.

The position of the paired wings is at any time an expression of the proton's temperature. When a proton gets hotter, it makes the proton a little longer, it expands. When it gets colder, its shape in Fig. 49b gets shorter, it shrinks as an iron rod will also do, or when it gets colder the proton will get a more square appearance, and when it is a part of some kind of substance that will do that, it may crystallize, like water and ice.

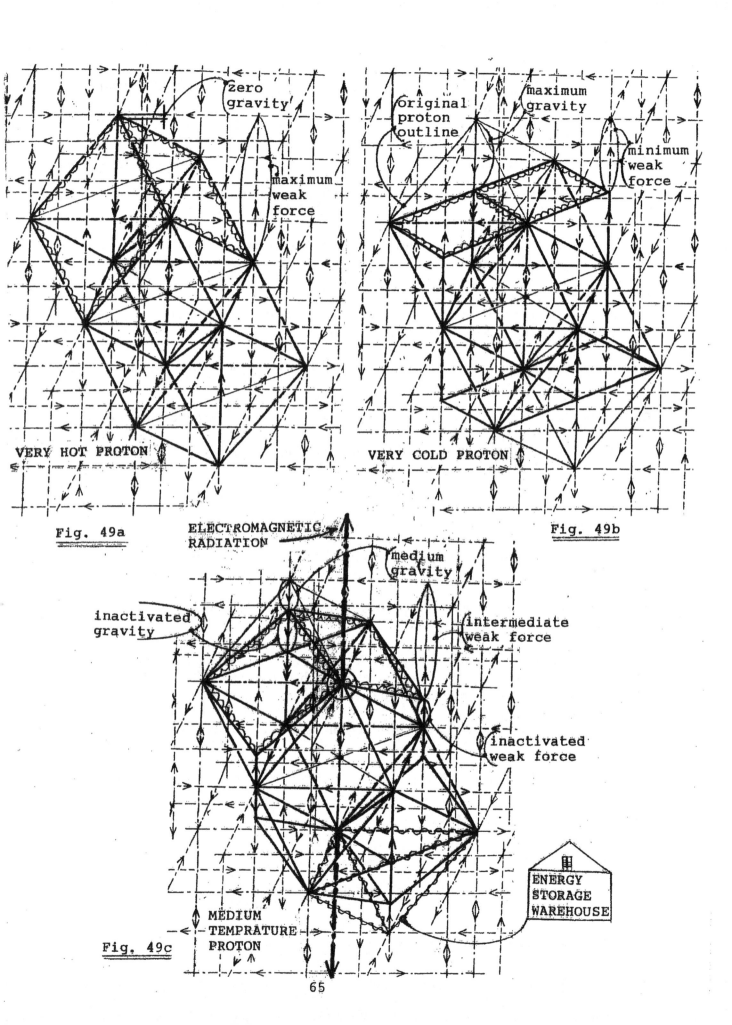

Fig. 49a

VERY HOT PROTON

zero gravity

maximum weak force

Fig. 49b

VERY COLD PROTON

original proton outline

maximum gravity

minimum weak force

ELECTROMAGNETIC RADIATION

Fig. 49c

MEDIUM TEMPRATURE PROTON

medium gravity

inactivated gravity

intermediate weak force

inactivated weak force

ENERGY STORAGE WAREHOUSE

65

The ultimate goal for its gravitational provision in the proton is its by Nature's assigned task to connect itself with other protons.

The drawings show that each proton has two locations for its gravity, being at opposite ends of the proton.

The gravitational strength of a proton can vary greatly, depending on its temperature or other factors.

The weak force will play a role in all of this with its inbred outbound force which, when pushing against the outbound force of another proton may succeed in certain instances in preventing two protons to make a connection.

Fig. 49d shows two protons approaching one another, with their externally located gravitational strength ready to connect with its opposing other proton.

The magnitude of their individual gravitational pull is determined by their actual variable structure, as explained in Fig. 49a,b,c.

In Fig. 49e the two protons are connected by their gravitational pull, which has caused the two protons to have their body enclosures to make contact. What has happened here is that each proton's 'hot-proton' outline, which represents temperature, has penetrated the opposing proton, feeling each other's temperature.

This intimate contact will instantly result in the two protons adjusting their own individual temperature to a temperature level that will be the same for both connecting protons.

What has just been described here in not chemical bonding, which will be described further on.

In summary it should be noted that the proton's gravitational force has a specific maximum, but that it can be reduced (in Fig. 49a) to zero.

The weak force has a built-in maximum in Fig. 49a, and its minimum is equal to half the length of the matrix's baseline in Fig. 49b. It can never be less than that, there will always be a certain amount of the weak force, pointing away from the particle.

This might suggest that this is the reason that the universe is always expanding because nothing ever gets lost in Nature.

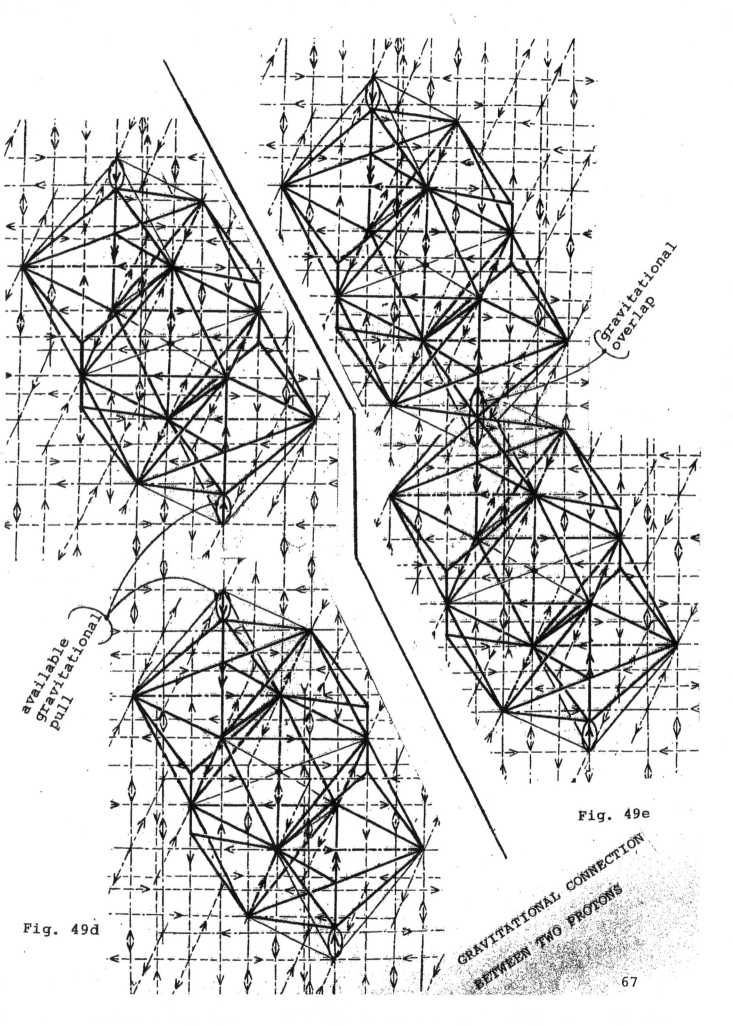

available gravitational pull

gravitational overlap

Fig. 49d

Fig. 49e

GRAVITATIONAL CONNECTION
BETWEEN TWO PROTONS

67

Simultaneous with the activity in gravity's corner, its weak force neighbor is also at work, continually emitting its repelling force, which like gravity reaches into infinity.

They are quarter pyramids 2E, 4S, 6N and 8W. See Fig. 50, shaded areas. Also shown here is the inward flow of gravity's pull, Fig. 51.

The weak force is usually said to have a very short range, in contrast with gravity that has an infinitely long pull.

However the weak force's pushing efforts do not get lost, in nature nothing gets lost ever.

Having made this statement, it is not much of a stretch to suggest that the pushing weak force and Einstein's Cosmological Constant are one and the same.

o o o o

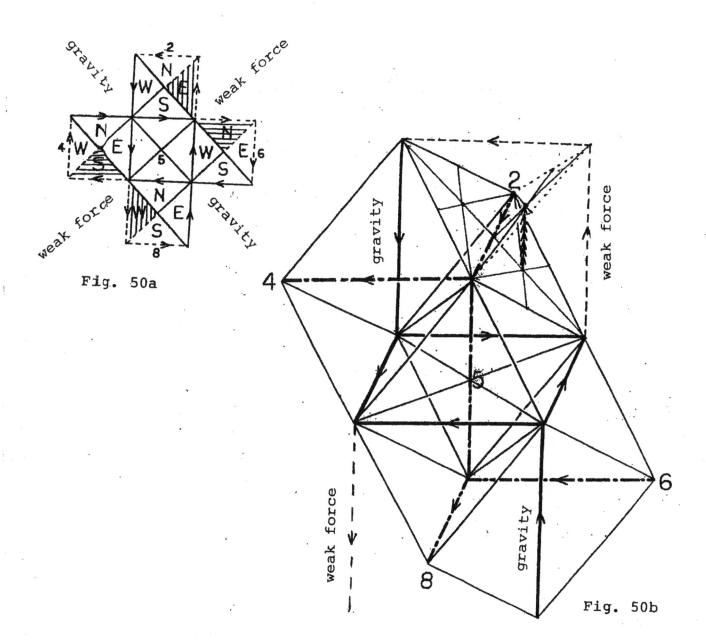

Fig. 50a

Fig. 50b

Newton's Third Law declares that each force has an equal and opposite reaction force.
However, we should not stop there:
When we read that law and we continue with its thought process, then we must conclude that this opposite force is only capable of delivering this opposing force when its own foothold behind it holds firm, but that foothold can only be firm if it has also its own opposing force behind it as well, to keep it in place and so on and on, and on and on, etc. etc.

This reaction force scenario applies not only to our muscles when we lift a weight, but also to the inner gravitational forces of the neutron, proton, as shown in Fig. 37a,b, where the composite 2-part structure of its central square provides the mechanism that will produce their nuclear mass.

The reach of all of the forces in the universe is immediate and infinite, there is no time lapse. If there were a time lapse, then Newton's "Equal and Opposite Force" statement of his 3rd Law would be incorrect.

Fig. 52a shows the direction of he pyramid cubes as they combine the folded over pyramids 2, 4, 5 and 8 as they squeeze captured Pyramid 5. Fig. 52b shows here an oversimplified drawing of the compression of Section 5 arrow cubes, which in the following chapter will explain how nuclear mass is created.

Fig. 53 is proof by means of a mechanical gear system that a force will transmit its presence instantly all the way, anywhere without delay.

This "without delay' statement means in this theory that the pull of gravity through space is instant, there would not be a "speed of gravity' that would be similar to the speed of light.

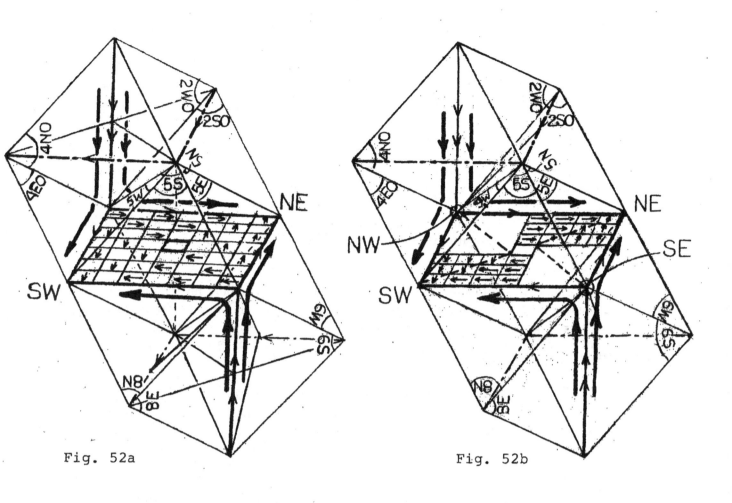

Fig. 52a

Fig. 52b

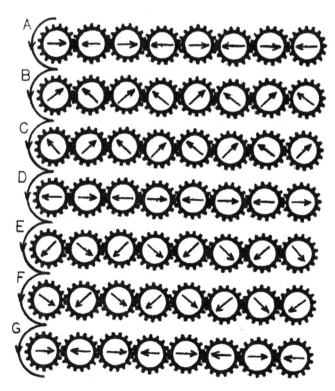

Fig. 53

A proton is used here as an example because it has a somewhat simpler structure then a neutron.

The folded over pyramids of Fig. 28, 29 have shown how the double pyramids have packaged themselves: Sections 2 and 4 folded up till they merged together as well as with the North and West pyramid slopes of central Section 5. Simultaneously, Sections 6 and 8 folded down till they merged together as well as with Section 5's East and South pyramid slopes.

Because the pyramids' internal structure consists of cubes with arrows that are separated from other cubes with arrows by cubes without arrows, throughout, it is possible to compress that "sponge like" assembly into a solid package. For example: when the upward thrust of the quarter pyramid portions of double pyramids 6 and 8 moves upward, it is then diverted towards the left and towards the rear by the quarter pyramid portions with the circled baselines of Sections 6 and 8.

The overall result of this joint effort is that in Fig. 54a the compressed bold-lined result is pushed into the center of the proton where it is met by its opposite identical twin that was produced by folded over Sections 2 and 4, in conjunction with Section 5. The bold-lined structure comprises exactly one third of the volume of the cube it inhabits. The length of the edges of the cubes are exactly half the length of the baseline and centerpost.

Fig. 54b and Fig. 54c are shown here on a smaller scale to save space. The two cubes on these drawings share a single cube at the center of this drawing, and it is that single cube that is the key to the nuclear mass of this particle, because it serves as the combined origin of the two opposing cubes. This paired combination provides for the super symmetry for the locked-in balance between all of the opposing components that contribute to the particle's structure.

All along this search, the double pyramid shapes of he various particles had dominated my thoughts that their nuclear mass would also have a similar, but much smaller configuration.

However, with all that trying, for nearly two decades, nothing worked.

SEE PAGE 184, CHAPTER: "CONSOLIDATING NUCLEAR MASS."
FOR MORE NUCLEAR MASS FORMATION DESCRIPTION.

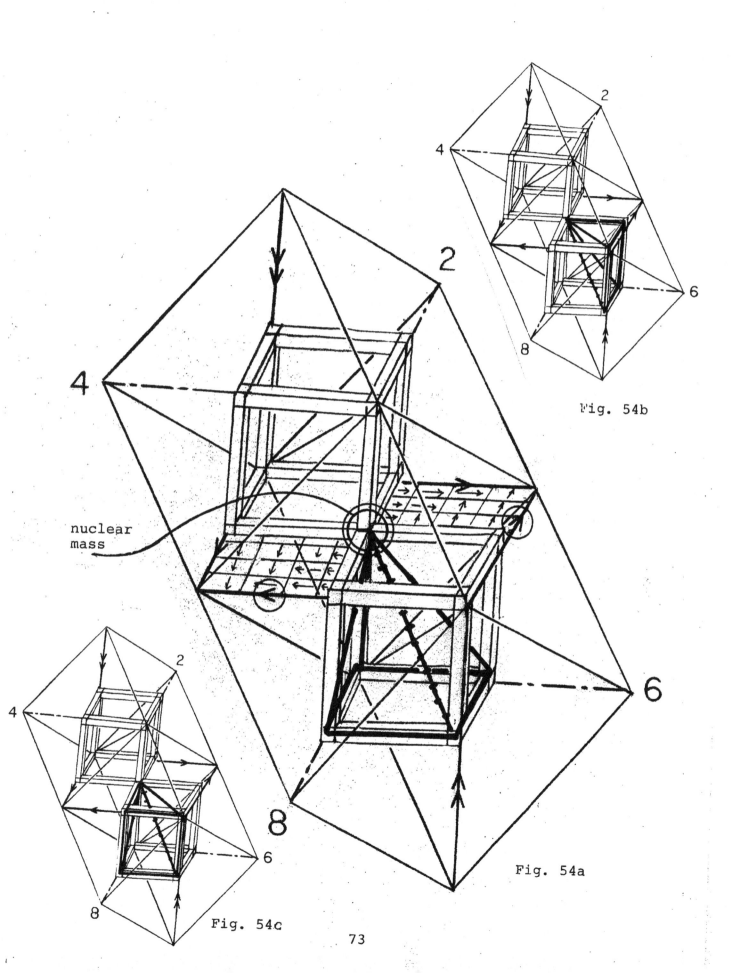

nuclear
mass

2

4

6

8

Fig. 54a

Fig. 54b

Fig. 54c

With my mind having been brainwashed for years and years by reading everywhere that the inward pull of gravity had produced spherical neutrons and protons and that the strong forces in and around their nucleus were of a swirling nature, it would just be silly to suggest that something so mundane as a cube that would be the keeper of a particle's nuclear mass.

Then, finally, on November 16, 2008, working on a blank piece of paper with a pencil that had a patient and cooperative eraser on its other end, I went for it, and broke free.

In my college years, right after World War II, we were taught that an electron's mass was set at 1,000 as a unit of measure, and that the proton was 1836 times heavier. I felt that what was needed now in this theory was an accounting system that provided a simple overview of the quantative relations between particles in the decimal system.

This theory with its arrowed cubes that are separated by open cubes without arrows provides an opportunity to establish a connection between the magnitude of the nuclear mass of a particle and the number of cubes in the pyramid shapes that keep them aligned in their fixed positions.

Because of the specific geometric images of this theory it is necessary to use the quantitative, relative measures between the particles that have been investigated here, meaning that the touchstone will be the electron with a nuclear mass of 1.0000.

Measured nuclear mass		Converted nuclear mass	
Electron	0.511 MeV	Electron	1.0000
Neutron	939.5731 MeV	Neutron	1838.6949
Proton	938.2792 MeV	Proton	1836.1636

At that 1.0000 basis, the neutron's converted value of 1838.6944 might be tested as to its suitability of fitting into a cube that has edges with a length L, so that the cube's volume = L^3 = 1838.6949. The answer was $L^3 = 12.25^3$ which actually = 1838.2656, which is very close: 1838.2656 divided by 1838.6949 = 0.9997665.

I noticed right away that $12.25 = (3\frac{1}{2})^2$, so that this equation can be written as
$$(3\frac{1}{2})^{2 \times} (3\frac{1}{2})^{2 \times} (3\frac{1}{2})^2 = (3\frac{1}{2})^6 = 1838.2656 - \text{neutron's nuclear mass.}$$

Keeping in mind that the neutron's mass number of 1838.6949 was based on the electron's measured MeV number of 0.511, which was probably rounded off from a slightly higher or lower number, we can then recalibrate this measured electron's MeV number with as its reference the neutron's calculated cubic number of 1838.2656, with the following equation:
$$\frac{\text{Measured Neutron 939.5731 MeV}}{\text{Calculated Neutron 1838.2656}} = \text{New Electron MeV} = 0.5111193.$$

The new proton's nuclear mass is then $\frac{938.2792}{0.5111193} = 1835.7342$ (A).

However, when we instead reference the calculation of the new proton number with the calculated new neutron number of 1838.2656, we then get the newly calibrated proton number as

New Neutron 1838.2656 x Old Proton 1836.1636 = New Proton = (B)
Old Neutron   1838.6949   1835.7348

Which one of the two, Proton A or Proton B, is the correct one? Both protons suggest that nature wants them to connect to a neutron, because their union would benefit both of them because their own fractured numbers, when combined, would make an exact whole number, being 3.674.0000. The actual additions are shown below:

Neutron	- 1838,2656		Neutron	- 1838.2656	
Proton A	- 1835.7342+		Proton B	- 1835.7348+	
Total	- 3673.9998		Total	- 3674.0004	

I suspect there is a slight proton MeV measurement error so that the proton's new nuclear mass might have been 1835.7344, which then, when coupled with neutron's 1838.2656 would make an even 3674.0000. Let us try: Suppose the old proton's MeV measurement should have been 938.2793 instead of 938.2792, then its new, converted nuclear mass would then be
983.2793 = 1835,7344. Bingo!
0.5111193

With this updated number for the proton, we can now take a look at Helium, which is made of 2 sets of neutron/proton combinations of their nuclear mass, as in a
Neutron - 1838.2656
Proton - 1835.7344
Total   - 3674.0000 times 2 = 7348.0000 for Helium.

What nature did here is have the neutron's fractional mass of .2656 dovetail perfectly with the proton's fractional mass of .7344, thereby providing fulfillment for both parties by creating a whole number without decimals. This reminds one of biology's double helix where its opposite components A and T and also C and G find structural fulfillment by their mergers.

In Helium's case it achieves fulfillment for its two "neutron with proton" mergers by not having any fractional component leftovers. And that, in combination with its perfect square shape and its perfect atomic weight of 4.0000 qualifies Helium as the building block of choice for the Periodic Table.

In the 1960's Murray Gell Mann and George Zweig proposed that all strongly interacting particles have quarks as a major part of their inner structure. Ever since there have been valiant efforts underway to isolate quarks as a separate component, but so far that search has not been successful.

Quarks are considered to be involved in the making/maintaining of nuclear mass. In general terms, the nuclear mass is located in the center of the proton and neutron, and the quarks are located around the nucleus.

It has been shown in the earlier part of these pages that the proton and the neutron both consist of two identical structures which are connected in their center, and the result is that each end of these particles has an identical Siamese twin at the other end, Fig. 56.

To keep the drawings simple and easier to understand, the quest for the quarks will be done at one end of the proton, which is an exact half proton, Fig. 57.

The wiggly lines on that half proton are dividing it into three geometrically identical pieces, which in their combined package resemble an umbrella. The wiggly lines are drawn along the edges where pyramid slopes meet. Each one third of the half proton has its own particular directions in which its arrows are flowing, each one third of the umbrella consists of one half of a pair of bottom to bottom pyramids which from a geometric point of view are two side by side tetrahedrons, Fig. 58.

It was found that these structures nicely fit into a system that can represent the quark theory, as shown on the following pages.

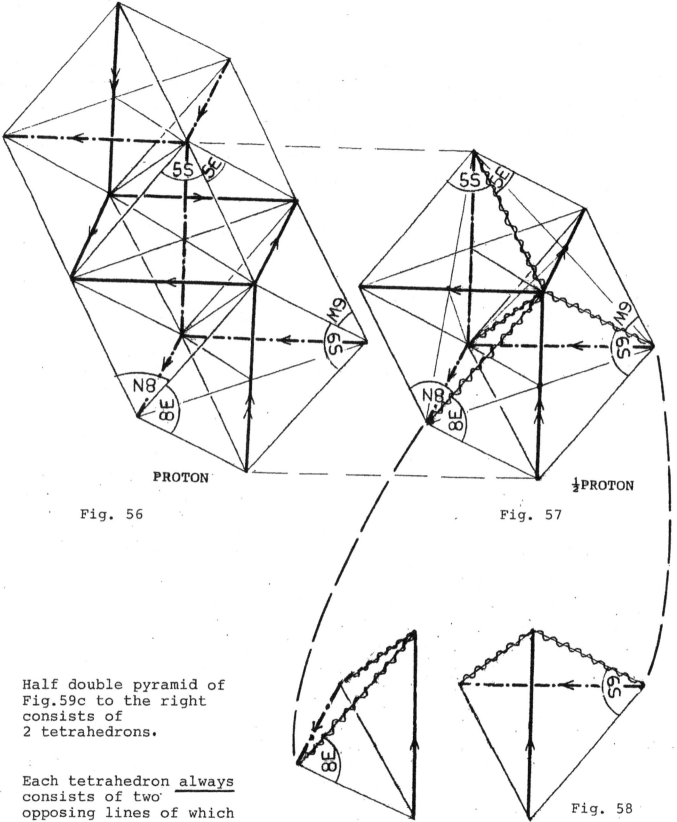

PROTON

Fig. 56

½PROTON

Fig. 57

Half double pyramid of
Fig.59c to the right
consists of
2 tetrahedrons.

Each tetrahedron always
consists of two
opposing lines of which
one is a baseline and the
other one is a centerpost.

Fig. 58

TWO TETRAHEDRONS.

Exploded view of
two half double pyramids.

77

In Fig. 59, half pyramid pair 5S/8N is being traversed by a common baseline, which departed from the umbrella's center, which is a point that has a positive charge. See Fig. 38 & 39.

In Fig. 60, half pyramid pair 5E/6W is being traversed by a common baseline, which departed form the umbrella's center, with its positive charge. See Fig. 38 & 39.

Each of the above half pyramid pairs occupies <u>one third</u> of the umbrella, their side by side attachments means that their combination represents <u>two thirds</u> of the umbrella. Adding to that that both half pyramid pairs are positive, it means that their quark number is $+\frac{2}{3}$.

Fig. 61a shows the third one third 6S/8E half pyramid pair, which has its central baseline arrow flowing TOWARDS the umbrella's center, which came from its negative base below. See Fig. 38 & 39.

Having reviewed all of this, we can conclude now that half pyramid pair 6S/8E is a one third part of the umbrella and that it is negative. However, we have seen in the chapter on gravity that half of the 6S/8E assembly was lost earlier for the construction of the neutrino, and that only half of that is left over, in Fig. 61b. This means that the one third part of the umbrella has now been reduced to half of that, which is one sixth of the umbrella.

It means that its quark number is $-\frac{1}{6}$.

The total quark count for this half proton is thus $+\frac{2}{3}-\frac{1}{6}$.

The whole proton quark count is double that

$$(+\frac{2}{3}-\frac{1}{6}) \quad + \quad (+\frac{2}{3}-\frac{1}{6}) \quad = \quad \frac{2}{3}+\frac{2}{3} \quad -\frac{1}{3} = \text{uud}.$$

Experiments have indicated that a quark is a pointlike particle. Let's examine Fig. 61a about that. In this theory the arrow in a baseline indicates the direction in which ALL arrows are flowing in any particular tetrahedron of which such baseline is a part. The wiggly-lined quark of Fig. 61a is therefore a <u>one-dimensional</u> object, because it does not have arrows that point in a second or third dimension. This quark's single direction arrow bundle therefore is 'pointlike.'

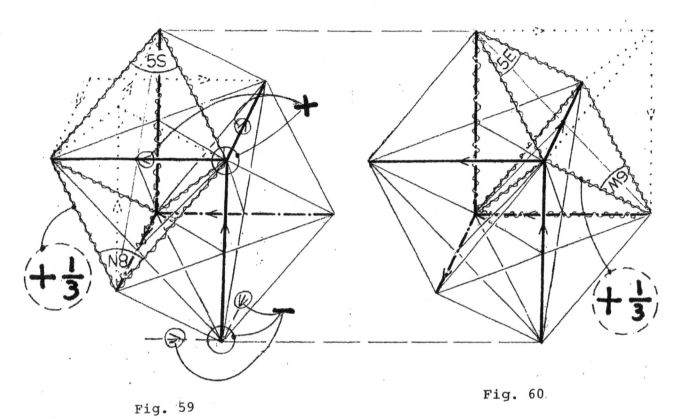

Fig. 59

Fig. 60

SEE Fig. 38,39

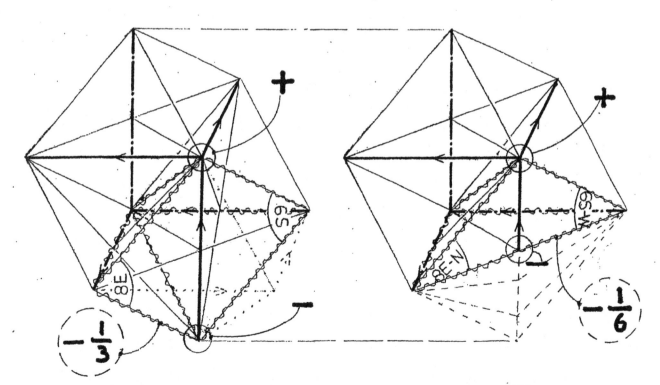

Fig. 61a

Fig. 61b

Having been reassured now that the quark count in the proton is correct, it is easier now to dissect the neutron. Starting with Fig. 62a, the $+\frac{1}{3}$ quark is the same as that of the proton, 5S/8N. Its _positive_ nature is due to the _positive_ corner from which its tetrahedron's baseline departs.

In Fig. 62b the neutron's structure now takes into account that it is still holding on to those components that later on may be released to be a part of the newly to be created negative electron. The wiggly lines of the neutron's 8S/8E portion delineate a tetrahedron and it is negative, because its "vertical" baseline departs from a _negative_ corner of the neutron. Its quark number is $-\frac{1}{3}$.

In section 6 there is an identical quark situation as just described for section 8.

We can then summarize this end of the neutron as

(Section 8): $+\frac{1}{3} - \frac{1}{3}$, with (Section 6): $+\frac{1}{3} - \frac{1}{3} = 0$.

However, the neutron's quark count is usually written as $+\frac{2}{3} - \frac{1}{3} - \frac{1}{3}$.

which has the same effect as $(+\frac{1}{3} + \frac{1}{3}) - \frac{1}{3} - \frac{1}{3}) = +\frac{2}{3} - \frac{1}{3} - \frac{1}{3} . = udd = 0$.

This is calculated for one end of the neutron only, but the same situation is present at the other end of the of the proton, thus doubling its number of quarks, but its overall neutrality is of course not affected.

In Fig. 62b and 63b the 8W has been marked with a square box around it. This is the arrow of the weak force, which was shown earlier in its connection with gravity.

Because this arrow points away from the neutron's center, it does not participate in the quark's activity of making nuclear mass. It is therefore not included in the quark count.

Fig. 63a shows how all three of the quarks at each of the two ends of the proton are packed together.

Fig. 63b, for the neutron, shows how all three of the quarks at its sections 8, 6, 4, and 2 are packed in a manner that renders the neutron electrically neutral at each of its four corners, which are still occupied by the potentially future half pyramid pairs which may take part to the making of an electron.

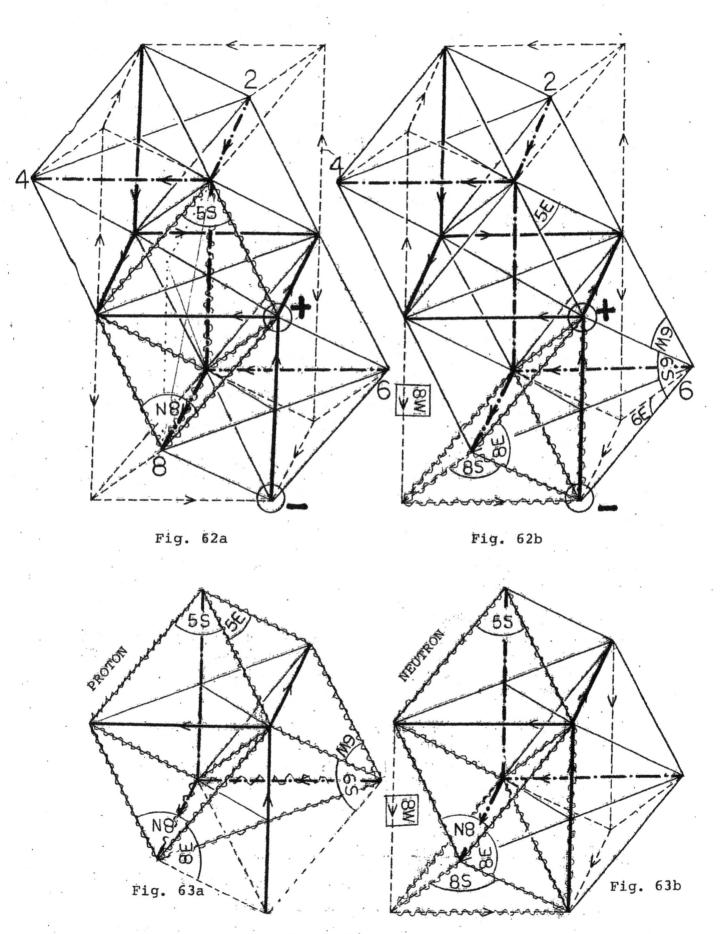

Fig. 62a

Fig. 62b

Fig. 63a

Fig. 63b

THE DEFINITION OF THE MATRIX:
THE OVERALL CONTROLLING CONCEPT OF ALL OF THE ALWAYS STRAIGHT LINES OF
THE MATRIX OF SPACE IS THAT ALL OF ITS ARROWS HAVE A REVERSED DIRECTION AT
THE OPPOSITE SIDE OF THE NODE WITH WHICH THEY INTERSECT.

Throughout this analysis it has become very clear that the inside out principle pervades this entire theory.
There is a continuous pushing and pulling between these two realms of existence, they both have similar
patterns of behavior, and their only way to survive and exist is to cooperate with each other.

We have seen this in all particles that were reviewed, and it can also be seen in the electromagnetic wave.
Fig. 64 is the same picture of the matrix that was already seen in Fig. 35.

In Fig. 64 the magnetic extension pattern has been highlighted with dash/dot lines as we already saw
it in Fig. 51, where folded over centerposts are inside out baselines of inside out pyramids that have as
centerposts the folded-over baselines of the original standard pyramids. The arrows in these pyramid
bases are all following each other in a continuous loop. However, folded over or not, baselines always
have arrows, centerposts are always vacant.

The electric charge loop pattern in Fig. 64 mimics that of the pyramid base of original square 5, having
pairs of arrows arriving or departing from base 5's corners. These loops also extend indefinitely, in the
same plane, in all of its own appropriate directions.

In these matrix patterns of magnetism and electric charge it is rather unexpected to see that at none of the
perpendicular intersections an arrow will continue in the direction that it came in, but that it had been
reversed.

At two centerpost's lengths above the just described directional patterns are planes with congruent patterns,
and they are also at two centerpost's length below it. They are all parallel with one another.

Midway between these parallel planes there are planes which all have arrows that are reversed from the
patterns described above. This has to do with nature's requirement that there be a balance in the inner
structure of particles, meaning that the inner balance needs to be structured in a manner that represents
an effort to return to zero with respect to its physical presence.

The electric and magnetic portions of the electro magnetic wave have a ninety degree angle between them
while they are weaving back and forth from one side of their centerline to the other. It is possible for the
wave to maintain this back and forth movement going because of the matrix's capability to transform
itself into the configuration that is needed at any moment.

Figure 65 shows how that works.

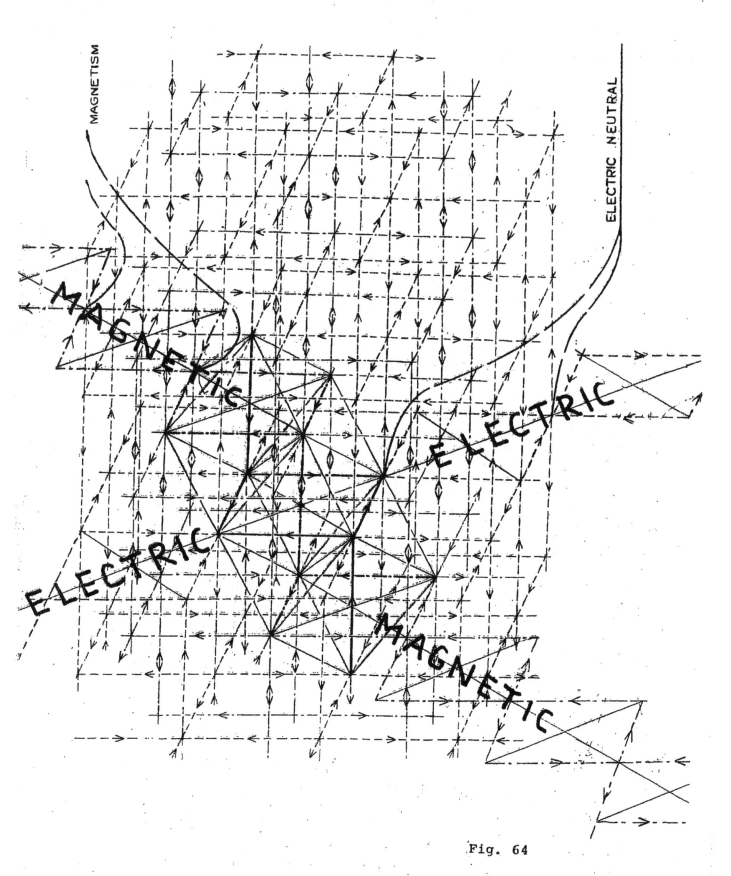

Fig. 64

An electromagnetic wave possesses a duality that makes it to be either a wave or a particle, the photon. With the help of the matrix as drawn in Fig. 65, and earlier in Fig. 35, it is possible to suggest how all of this might be happening:

The proton at the bottom of the page starts it all when it wants to release some of its energy in the form of heat into its environment.
(In order to facilitate this explanation, let us assume that the proton's square 5 is positioned in a horizontal plane, and that the electromagnetic wave emits form that in a vertical direction.)

In the chapter about being positive or negative, in Fig. 39 it was shown that it requires three baselines (and no more and no less), at a cube's corner to produce an electric charge. The proton drawings show that the proton's central section 5 has only two baselines meeting at its NE and SW corners. This means that those corners are neutral, not positive, not negative, which makes those corners an ideal location for transmitting impartially the electrical message and energy to the outside world.
It will thereby use the ever-present matrix of open space to transfer this electric message as needed. The matrix is the carrier, it is always there available and able, forever.

Depending on the magnitude of the transmitted energy, the amplitude and wavelength will vary, but at the receiving end of the journey it will deliver the energy package.

In the first step of the electric wave, the proton's square 5 extends itself in a NE direction, where it then also extends itself upward. After a distance of the length of a centerpost upward, it connects with a square that has arrows which are all opposite to the arrows of square 5. When it goes up another centerpost distance, it then finds a square that is congruent with square 5.

At this point the magnetic portion at the upper NW corner of the proton has kept pace with the proton's electric wave portion at square 5's NE corner. Square 5 is the location within the proton that contains its nuclear mass, and it is therefore most likely that that relationship is responsible for the observance that the wave at its nodes where the electric and magnetic wave meet transforms form a wave into a photon particle.

When the electric wave and the magnetic wave intersect on their joint effort's way, they will for a very short moment recreate an almost phantom structure of their proton's central section 5. The result is a particle-like structure, with just energy. This meeting place or node is the photon.

The series of inside out pyramids of the magnetic wave of Fig. 65 have a gravity baseline as their centerpost, which probably accounts for the similarity of the formulas of attraction for gravity and magnetism. It may be this gravitational component that probably causes the light of a distant star on its way to an observer on earth to be deflected by our sun's gravitational pull, as Einstein had predicted.

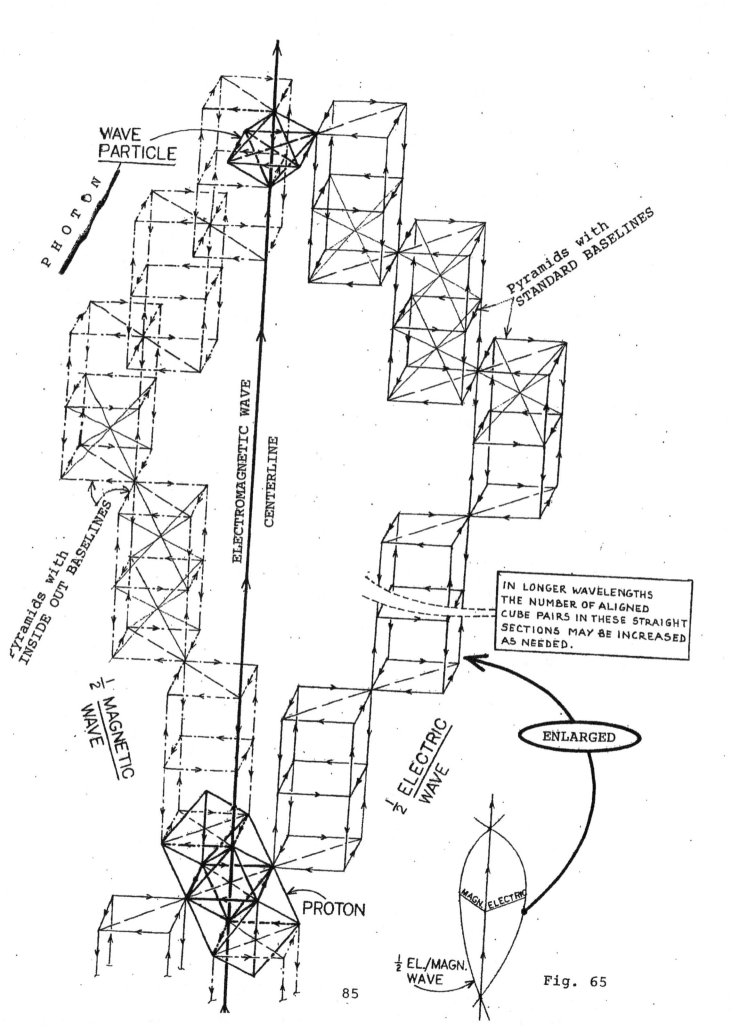

WAVE
PARTICLE

PHOTON

Pyramids with
STANDARD BASELINES

ELECTROMAGNETIC WAVE
CENTERLINE

Pyramids with
INSIDE OUT BASELINES

IN LONGER WAVELENGTHS
THE NUMBER OF ALIGNED
CUBE PAIRS IN THESE STRAIGHT
SECTIONS MAY BE INCREASED
AS NEEDED.

$\frac{1}{2}$
MAGNETIC
WAVE

$\frac{1}{2}$ ELECTRIC
WAVE

ENLARGED

PROTON

MAGN. ELECTRIC

$\frac{1}{2}$ EL./MAGN.
WAVE

Fig. 65

85

The proton is known to be a very stable particle, which distinguishes itself from other particles by having the largest number of characteristics that can be observed in many ways.

Fig. 66a has all of those characteristics collected in one drawing that shows their individual relationships as well as the number of locations where each one of them can be found.

The dashed lines at the exterior of the proton belong to the neutron's extremities which in a future break up of the neutron will become components of the electron, thereby revealing the proton's later efforts to recreate the neutron, attempting to return to the womb.

It is the electromagnetic radiation that carries within its structure more traits of the proton than any other particle.

THE PROTON'S INNER STRUCTURE.

In its most basic form it consists of an integrated assembly of tetrahedrons. Each of these tetrahedrons consists of a (solid lined) baseline that is connected with its ends by means of the corner ridge of two intersecting pyramid slopes to the ends of a (dash/dot lined) centerpost, Fig. 66b. The solid lined baseline represents a "standard" pyramid's presence, the dash/dot line represents the baseline of an inside out pyramid.

This means that each tetrahedron has one leg in the "standard" world, other leg stands in the "inside out" world,

NEITHER ONE CAN EXIST WITHOUT THE OTHER.

A quark consists of two tetrahedrons which share a baseline.
A quark with a shared standard baseline is positive, Fig. 66c.
A quark with a shared inside out baseline is negative, Fig. 66d.

Experiments which try to tear quarks apart will give rise to the strong force, which will only exist when outside forces will try to separate a tetrahedron's standard baseline from its inseparable companion, the inside out baseline.

The elasticity of the quarks' mutual bond between each other is nothing less than nature's need to preserve an unbreakable bond between its two equivalent components of the two sides of reality.

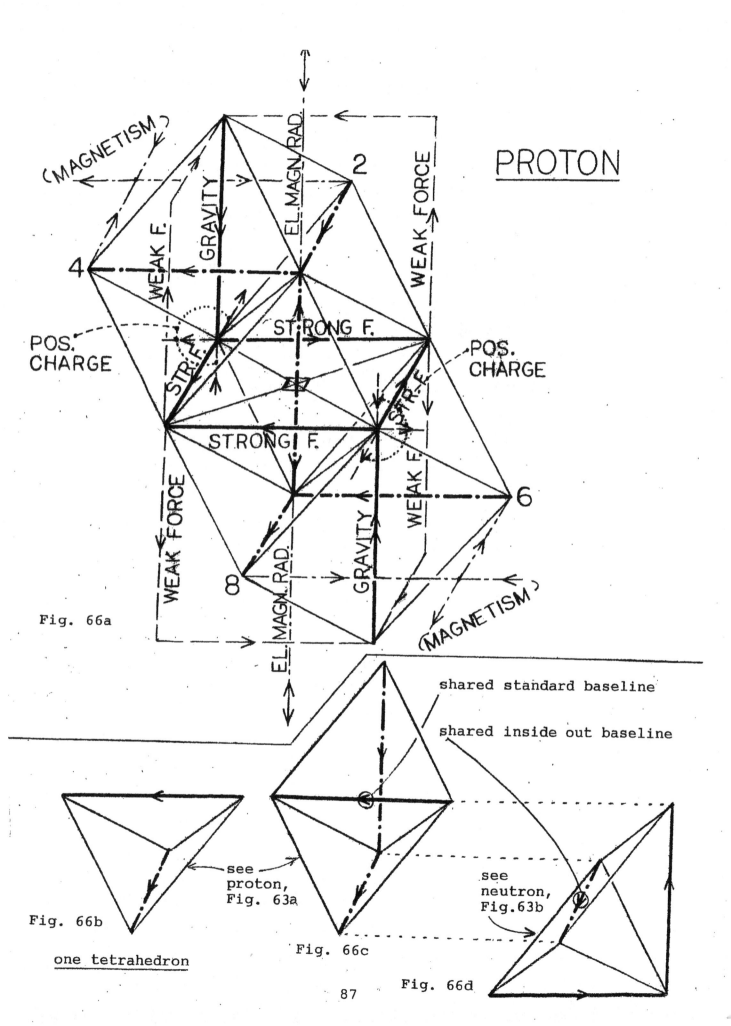

PROTON

MAGNETISM

GRAVITY

EL. MAGN. RAD.

WEAK FORCE

2

WEAK F.

4

STRONG F.

POS.
CHARGE

STR. F.

STR. F.

POS.
CHARGE

STRONG F.

WEAK FORCE

WEAK F.

6

8

GRAVITY

EL. MAGN. RAD.

MAGNETISM

Fig. 66a

shared standard baseline

shared inside out baseline

see
proton,
Fig. 63a

see
neutron,
Fig.63b

Fig. 66b

one tetrahedron

Fig. 66c

Fig. 66d

Fig. 67a is a slightly modified drawing of Fig. 41b of an electron that had lost its lower half for the benefit of the neutrino. All pyramid slopes are now identified, and its square dash-dot magnetic 'circle' with its CW rotation has been shown rounded off, because it is perceived like that.

Fig. 67b is a simplified rendition of Fig. 67a, it is easier to comprehend this way, its four downward pointing centrally located baselines show now a single line with four arrows in series.

Fig. 67c is the simplest drawing of all. Rudiments of its cross with its four converging arrows are still shown, the cross is actually not there anymore, except for inside the single cube that is locked in at its top center, see next Fig. 68a,b,c. This means that these four converging baselines with their attached quarter double pyramids have shrunk and fallen away. They were 2E/2EO, 4S/4SO, 6N/6NO and 8W/8WO.

What remains in a pyramid that starts with a single cube on top and that may grow in Fig. 67d to a full size assembly of the upper half of quarter double pyramid sections 2N/2NO, 4W/4WO, 6E/6EO and 8S/8SO.

It is important to notice that all of the arrows of this pyramid are pointing downward, except that single cube at the top, as shown in Fig. 41f.

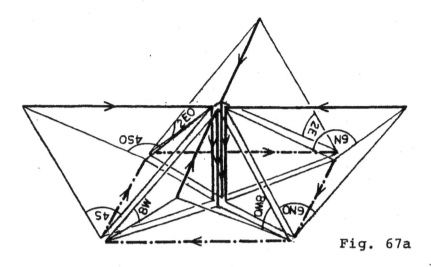

Fig. 67a

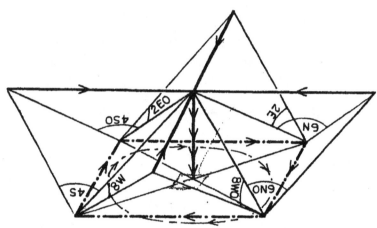

Fig. 67b

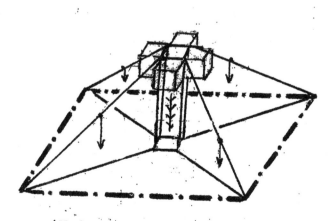

(Not to scale)

Fig. 67d

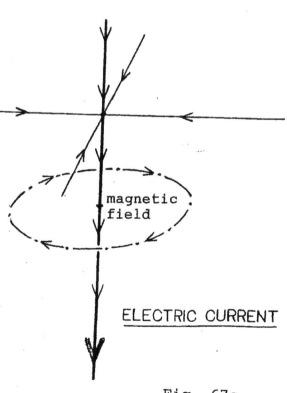

magnetic
field

ELECTRIC CURRENT

Fig. 67c

In reviewing the electron, we are now only concerned with the downward pointing cubes which are contained within the confines of the bold-lined pyramid on Fig. 68a. In Fig. 68b it shows how the four colliding converging arrows meet in the top center cube, as shown in the Fig. 68c detail.

It is right there that two pairs of opposing force arrows stop each other, which creates nuclear mass. It is similar to the mass-making scenario of the proton. Newton's third law fits in here nicely: When my fist hits a brick wall with a considerable force, then I find out that it is the inertia of that wall's mass that delivers the reaction force. That's why that single cube of Fig. 68c is the electron's rest mass.

When an electron gets energized, experiments have shown that it may accumulate a mass that may be as high as about 900 mass units, which is about half of the proton's mass of 1836 mass units. This added mass of the electron can now only be generated by all of those downward pointing speeding force arrows of a highly accelerated electron which potentially can be energized within the confines of the bold-lined, pyramid of Fig. 68a. Remembering Newton, we see in Fig. 68a that an electron with its rest mass of one is in a pristine matrix setting that actually represents its inescapable matrix which mandates that the bold-lined pyramid be filled with downward pointing forces which will be reacted to by the same number of upward pointing forces: Action equals Reaction. The circled electron sites all have of course the same matrix, see Fig. 68e.

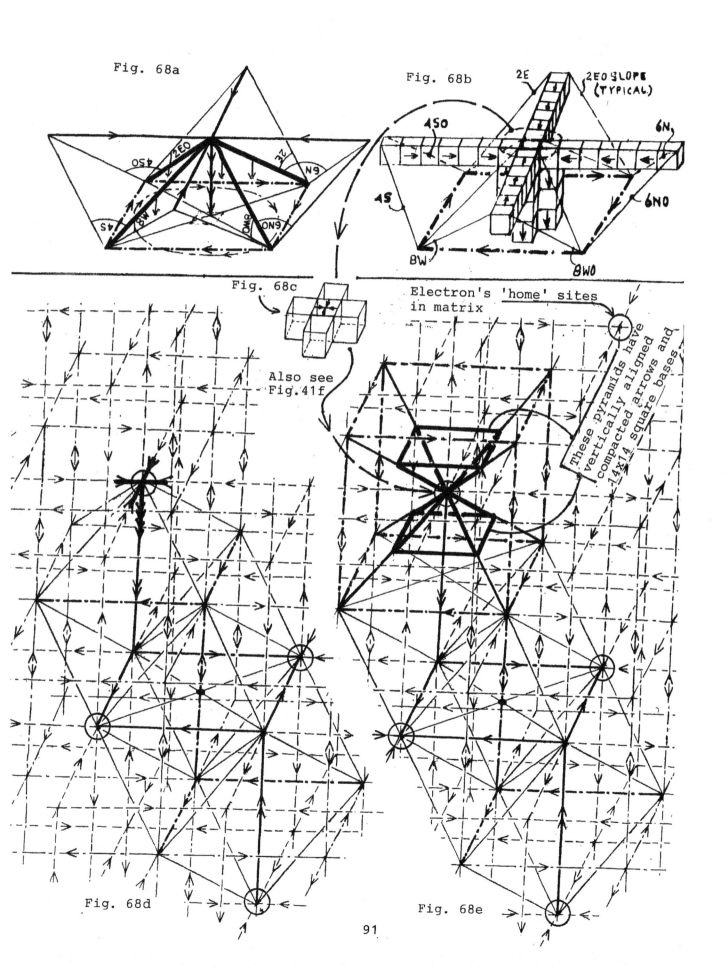

Fig. 68a

Fig. 68b

2E 2EO SLOPE (TYPICAL)

4SO 6N,

4S 6NO

8W 8WO

Fig. 68c

Also see Fig.41f

Electron's 'home' sites in matrix

These pyramids have vertically aligned compacted arrows and 14x14 square bases

Fig. 68d

Fig. 68e

91

Next: With some trial and error we can find out where a heavy electron keeps its 900 mass units. A proton's two mass packages are anchored alongside the proton's central centerpost in a symmetrical and centrally located way. It has mass at the center of an inside out cube that is made of centerposts of standard pyramids made with standard baselines.

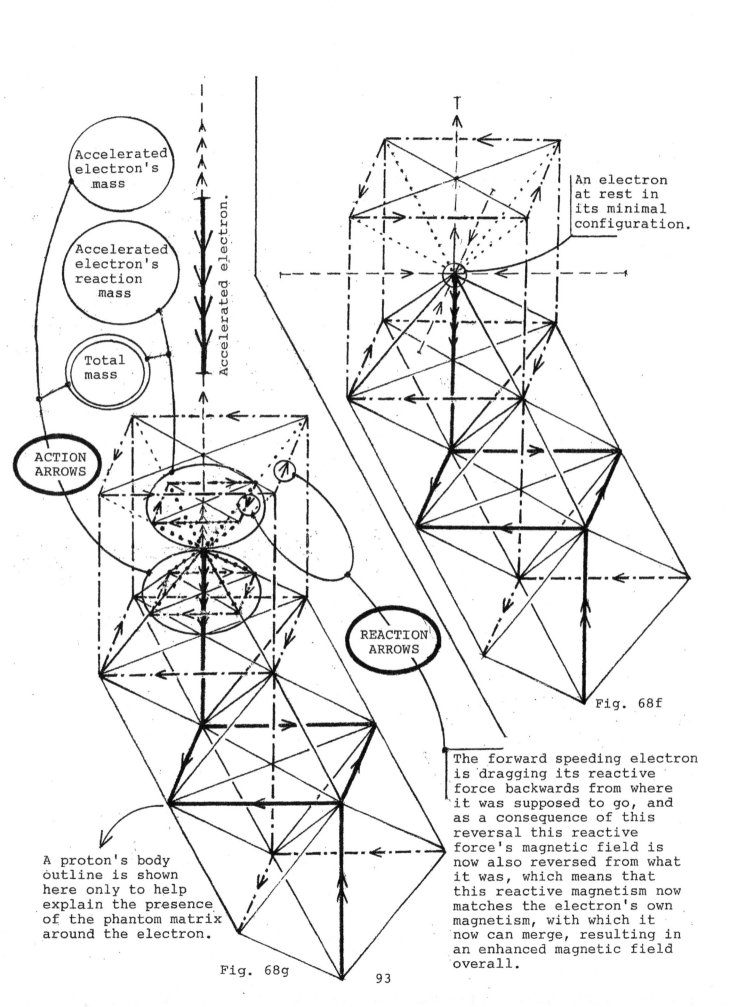

Accelerated electron's mass

Accelerated electron's reaction mass

Total mass

Accelerated electron.

ACTION ARROWS

REACTION ARROWS

An electron at rest in its minimal configuration.

Fig. 68f

A proton's body outline is shown here only to help explain the presence of the phantom matrix around the electron.

The forward speeding electron is dragging its reactive force backwards from where it was supposed to go, and as a consequence of this reversal this reactive force's magnetic field is now also reversed from what it was, which means that this reactive magnetism now matches the electron's own magnetism, with which it now can merge, resulting in an enhanced magnetic field overall.

Fig. 68g

93

Borrowing some words form Brian Greene's "The Elegant Universe" in which its subsection 'Spin' describes the spin of an electron as:

'Being not a transitory state of motion for more familiar objects that, for some reason or other, happen to be spinning. Instead, the spin of an electron is an INTRINSIC property, much like its mass or its electric charge.'

The spins of various particles are summarized as follows:

SPIN-1/2 ....   All of the matter particles, proton, neutron, have spin equal to that of the electron.

SPIN-1 ....   Non-gravitational force carriers, such as photons, weak gauge bosons, gluons.

SPIN-2 ....   The spin or the hypothesized graviton.

The pages of this manuscript "Decoding the Periodic Table" describes a theory of interacting forces, and on close scrutiny it fits perfectly in the above described classification of various spins. It was found that the measure of spin is intimately associated with these particles' internal TETRAHEDRON structures as shown earlier in quarks Fig. 62a, proton structure Fig. 66a,b,c,d and here again in an electron as Fig. 69a.

In this theory: Spin is measured as baseline's length versus its associated tetrahedron's centerpost's length.

Fig. 69a. The electron has only half as much baseline at its inside-out location as the full length centerpost has at its inside-out location on the outside, for a SPIN - ½.

Fig.69b. The non-gravitational force carriers such as photons, weak gauge bosons and gluons have a full size baseline as well as a full size centerpost, which gives them a SPIN - 1.

Fig. 69c. The hypothesized graviton is made by 2 full size side by side baselines which each have their own centerpost, for a SPIN - 2.

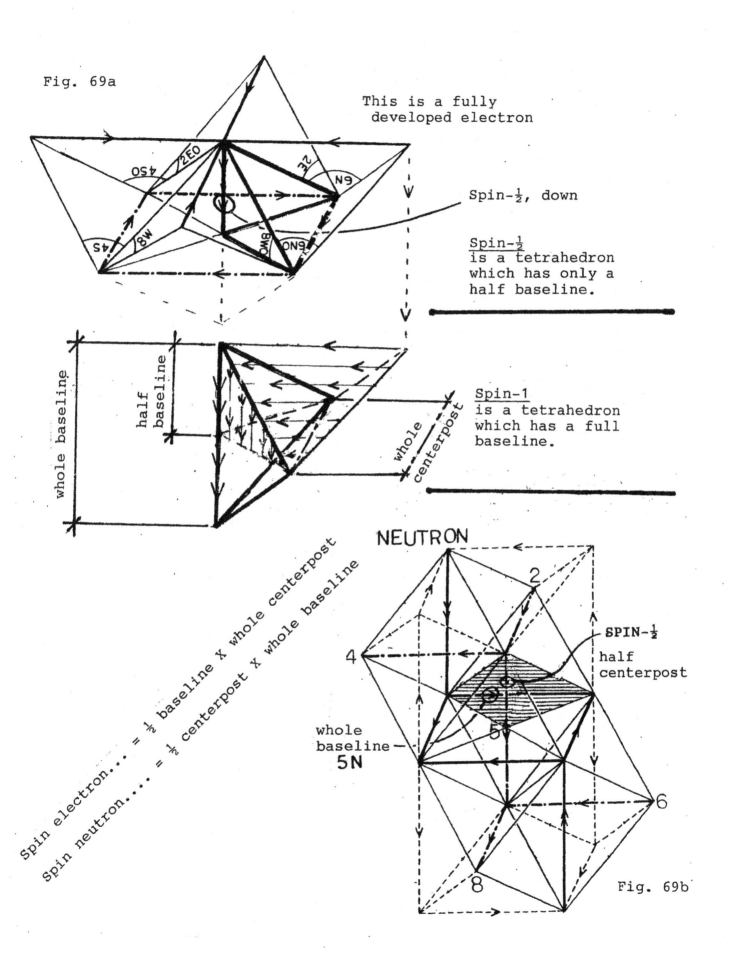

Fig. 69a

This is a fully
developed electron

Spin-½, down

Spin-½
is a tetrahedron
which has only a
half baseline.

Spin-1
is a tetrahedron
which has a full
baseline.

whole baseline

half baseline

whole centerpost

NEUTRON

SPIN-½
half
centerpost

whole
baseline—
5N

Spin electron.... = ½ baseline X whole centerpost
Spin neutron.... = ½ centerpost X whole baseline

Fig. 69b

Knowing the spin values of 1/2, 1 and 2, coupled with the description of the various particles where they belong in any of these three classification, provides us now with the tools to locate their respective intrinsic spin in those particles.

Throughout this theory it has been pointed out that the arrow in a pyramid baseline is an indication that:

> All of the arrows which are contained within the tetrahedron of a quarter of a standard bottom-to-bottom pyramid pair with its baseline and a centerpost are all pointing in the same direction as that arrow in a baseline.

To hear it easier to explain, we'll start with
...... SPIN-1, Fig. 69d, (non-gravitational force carriers, such as photons, weak gauge bosons, gluons.)

We know from Fig. 66a where the weak force is located, and Fig. 69c shows how the entire shaded quarter double pyramid which belongs to Section 2's 2E/2EO is a whole tetrahedron.

...... SPIN-1/2: Fig. 69e (mass particles)

The tetrahedron which is involved here is only one half of the 5N/5NO tetrahedron of section 5 as far as the proton or neutron are concerned. In the exploded view of the earlier shown electron of Fig. 69a, we see the half tetrahedron with Spin-1/2.

......SPIN-2, Fig. 69f (graviton)

This is supposed to be the spin of the hypothesized graviton. From Fig. 66a we know where gravity is located, and its double arrow in the combined baselines represent the attached full tetrahedrons of section 2's 2W/2W) (shaded) tetrahedron and also section 4's 2N/21NO tetrahedron.

Those arrow-filled tetrahedrons operate as a homogenous bundle of arrows which for a large part categorize a particle's characteristics and behavior.

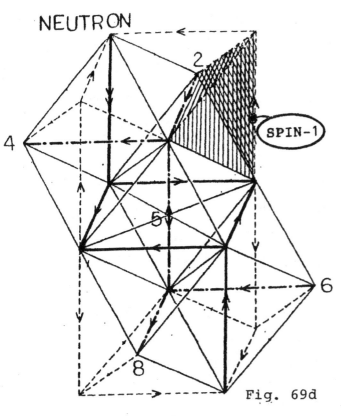

NEUTRON

SPIN-1

Fig. 69d

Spin-1...One whole tetrahedron

Non-gravitational force carriers

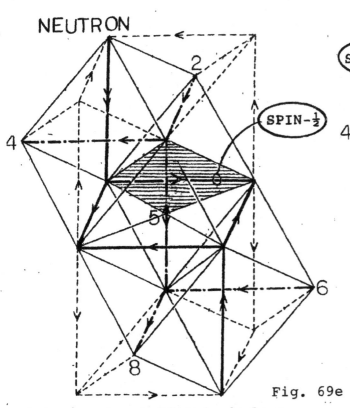

NEUTRON

SPIN-½

Fig. 69e

Spin-½...One half tetrahedron

All matter particles

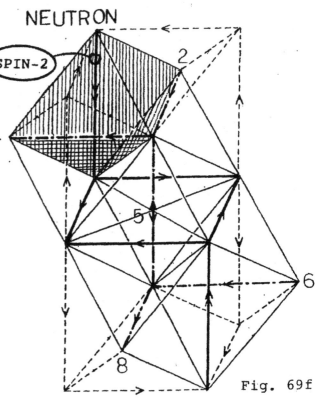

NEUTRON

SPIN-2

Fig. 69f

Spin-2... Two whole tetrahedrons

Hypothesized massless graviton

In the discussion about the electron in an earlier chapter it was already clear in the right half of Fig. 42a that a half electron has the same physical structure as one half of a proton, except that all arrows of the half electron are exactly the reverse of their physically corresponding arrows on the half proton.

Taking it a step further, it can be shown in this theory that when both halves of the proton are substituted for two half electrons that we are then looking at a negative proton.

This negative proton however does not have any nuclear mass in its center, because the circled compressed cubes in, for example, the SE corner of its central section 5 bleed out through the double circled leak.

For obvious reasons the paired half electron structure that was described above can be named an ANTIPROTON.

In Fig. 35 it was shown earlier how the proton is surrounded by an extensive spatial matrix that I was able to deduct from the proton's and neutron's baselines and centerposts' intertwined relationship.

This matrix envelopes EVERY particle there is over a short range, and when two particles join each other and make a firm connection, then this can only take place if the matrix from the one blends in perfectly with the matrix of the other. Nature is very precise. Examples of this will be shown later with chemical bonding and the structure of helium.

Figure 70 was a surprise that I discovered late in this analysis when I was on my umpteenth effort to figure out the orbit of the electron.

As the drawing shows, the proton is flanked and connected above and below by a negative proton, each of them are structured like a proton, but with their arrows opposite to those of the proton.

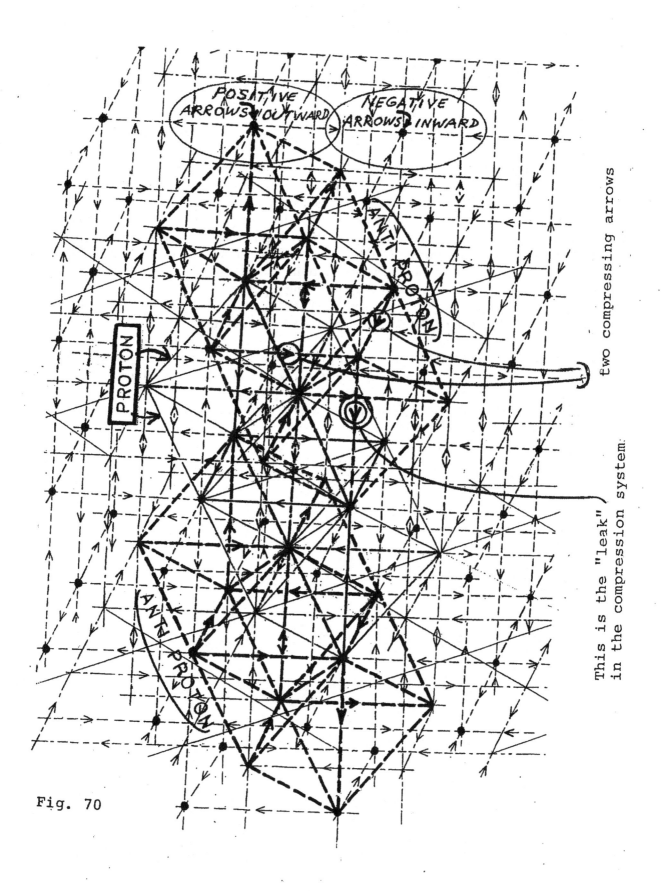

POSITIVE ARROWS OUTWARD

NEGATIVE ARROWS INWARD

ANTI PROTON

PROTON

ANTI PROTON

This is the "leak" in the compression system.

two compressing arrows

Fig. 70

99

WE need to go back now in this theory to Fig. 41b,c,d, of which Fig. 41d, shown here as Fig. 71, represents an electron in its smallest configuration, which is the one in which it orbits around a proton. It is known from experiments that subject to an electron's energy it possesses at one point, it may jump up or down to a higher or lower orbit. Orbital changes in this theory are destined to obey a path of fixed stationary locations.

The matrix concept in this theory has an answer for that. Earlier shown Fig. 38 shows the two positive locations on a proton. Fig. 39 is a neutron which shows how its two negative charge locations neutralizes the proton's positive charges, shown here as Fig. 72. Comparing now this proton/neutron combination with the knowledge that this package with its fixed structure is surrounded by a matrix that actually dictates the configuration and behavior of these particles, then we can use Fig. 39 here, reprinted here as Fig. 73, where we can use a proton, surrounded by its matrix, to spot all of those negative locations around a proton that are custom fitted to receive a small electron, as is shown on the following pages.

The observed orbit of an electron around a proton is a deception.

What it is, in this theory, is that an electron with its intrinsic spin has a choice to occupy one of the several stationary positions in the gossamer matrix that surrounds the proton.

The intrinsic spin of the electron seems to observers to be an imitation of our moon orbiting our earth, or our earth and other planets circling the sun.

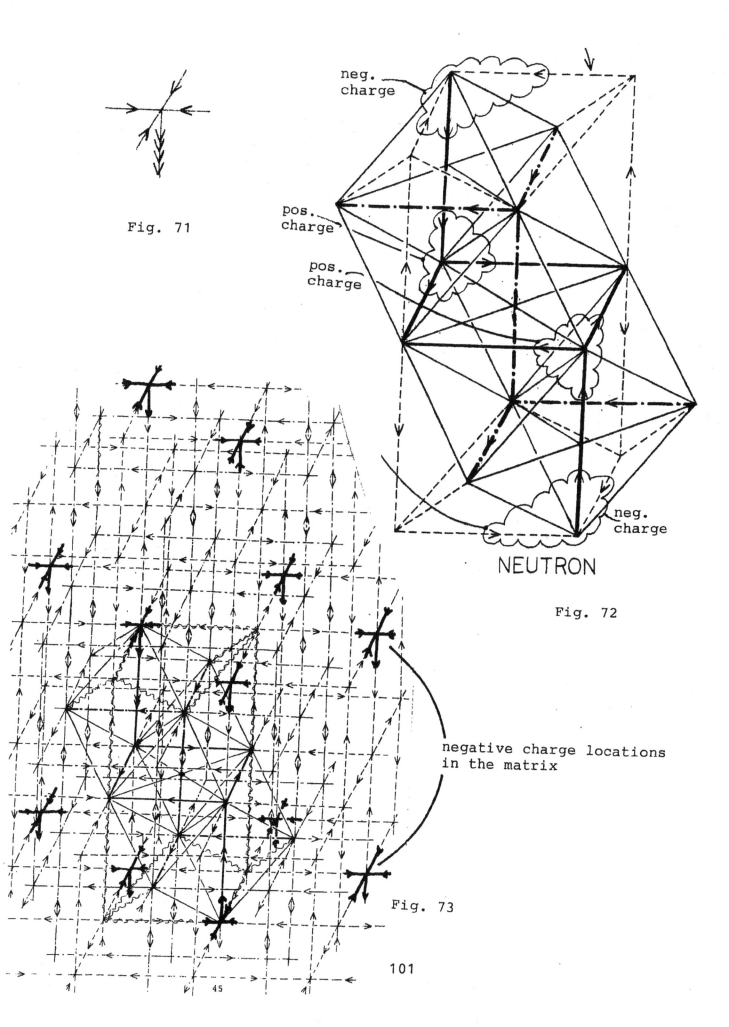

Fig. 71

neg.
charge

pos.
charge

pos.
charge

neg.
charge

NEUTRON

Fig. 72

negative charge locations
in the matrix

Fig. 73

101

45

The most surprising feature of the matrix pattern is the fact that none of the straight lines' force arrows continue in the same direction in which they were going after they have gone past an intersection. Without exception, the arrow's direction has been reversed when they come out at the other side of a node.

In this theory: The entire assembly which lives in this matrix is the ultimate expression of Newton's 3rd Law which states that every force has an equal and opposite reaction force.

Everyone agrees with that statement, but one sees it rarely quoted in today's physics.

In the world of Physics, Newton's 3rd Law is controlling the universe, all other laws are derivatives of that.

Also, none of the matrix's lines shows even the slightest hint of being curved someplace, and that feature in particular makes it for nature very much easier to construct and deconstruct particles and then use the remains to construct other particles. Nature is the ultimate recycler.

Fig. 74 shows in bold lines a proton as it fits in its surrounding matrix. At two locations there is a schematic drawing of the electron outside the proton. The dotted lines around the proton represents the actual path that the electron travels around the proton, while it may stop and stay for awhile at each one of four suitable locations in the matrix that exactly match the electron's own arrow pattern, shown here with a small cloud around it. Coupled with the perceived intrinsic spin of the electron, it will appear to an observer as if an electron is actually traveling at a very high rate of speed around a proton.

An electron's high or low orbit is determined by the stationary position in the matrix it occupies.

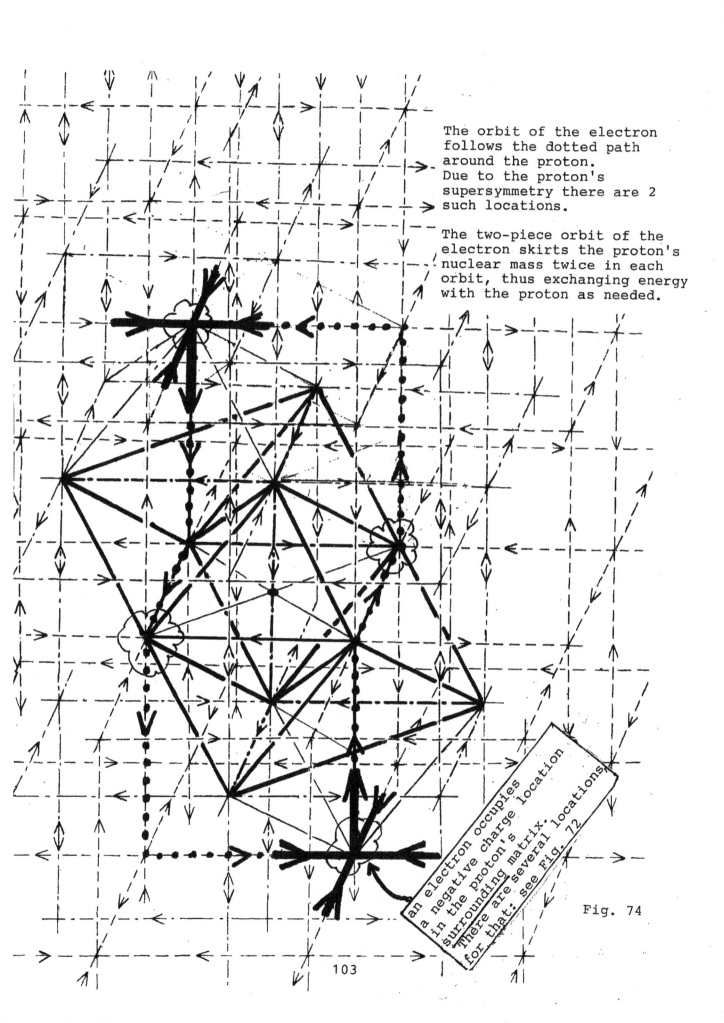

The orbit of the electron
follows the dotted path
around the proton.
Due to the proton's
supersymmetry there are 2
such locations.

The two-piece orbit of the
electron skirts the proton's
nuclear mass twice in each
orbit, thus exchanging energy
with the proton as needed.

an electron occupies
a negative charge location
in the proton's
surrounding matrix.
There are several locations
for that: see Fig. 72.

Fig. 74

103

A nice illustration in the November 2006 issue of Scientific American, named "The Dark Ages of the Universe," by Abraham Loeb, describes how an electron that has a spin that is opposite to the spin of the proton that it orbits, can be flipped 180 degrees when it gets hit by a photon, Fig. 75. The photon gets then absorbed, and the electron spin is now in the same direction as the spin of the proton.

When thereafter the captured photon gets released then the electron flips back to its original orientation, which its spin opposite to the spin of its proton.

Fig. 76 shows this electron with a down spin that is 180 degrees opposite to the spin of the proton directly above it. Fig. 77 shows that same electron flipped over to the other side of the node, now matching the proton's spin. It is the pre-ordained matrix pattern with its opposing arrows at opposite sides of this intersection of pyramid baselines that is capable of providing an immediate flip-over, literally on the same spot. This suggests that the electron's energy has moved form one side of the node to the other side of that node because of a better fitting energy situation.

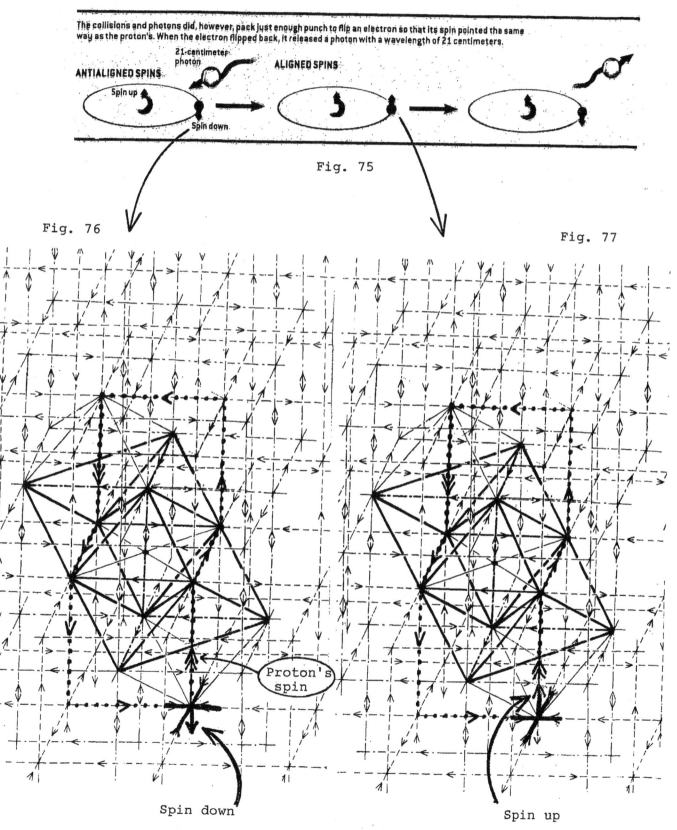

The collisions and photons did, however, pack just enough punch to flip an electron so that its spin pointed the same way as the proton's. When the electron flipped back, it released a photon with a wavelength of 21 centimeters.

ANTIALIGNED SPINS

21-centimeter photon

ALIGNED SPINS

Spin up

Spin down

Fig. 75

Fig. 76

Fig. 77

Proton's spin

Spin down

Spin up

105

When two particles make a chemical bond we may surmise that both particles are getting a type of fulfillment out of that union by sharing a useful structural portion of its own with a useful structural portion of the other particle in order to achieve a higher level of completeness for both participating parcels.

With that in mind we can analyze a proton's structure with the knowledge that a proton is actually a crippled neutron that at four locations had lost two half bottom-to-bottom pyramids to the precursor of the electron, see Fig. 78.

The pattern of the matrix that supports this theory has all along been the first foundation under the various characteristics and shapes of its particles.

Now, on the subject of chemical bonding it is the non-relenting pattern of the matrix that completely controls how and where on the proton chemical bonding takes place.

Fig. 79 shows side by side the 'front' and 'back' of a proton, its four shaded diamond shaped areas mark the scars that remind us of the neutron's lost electron components.

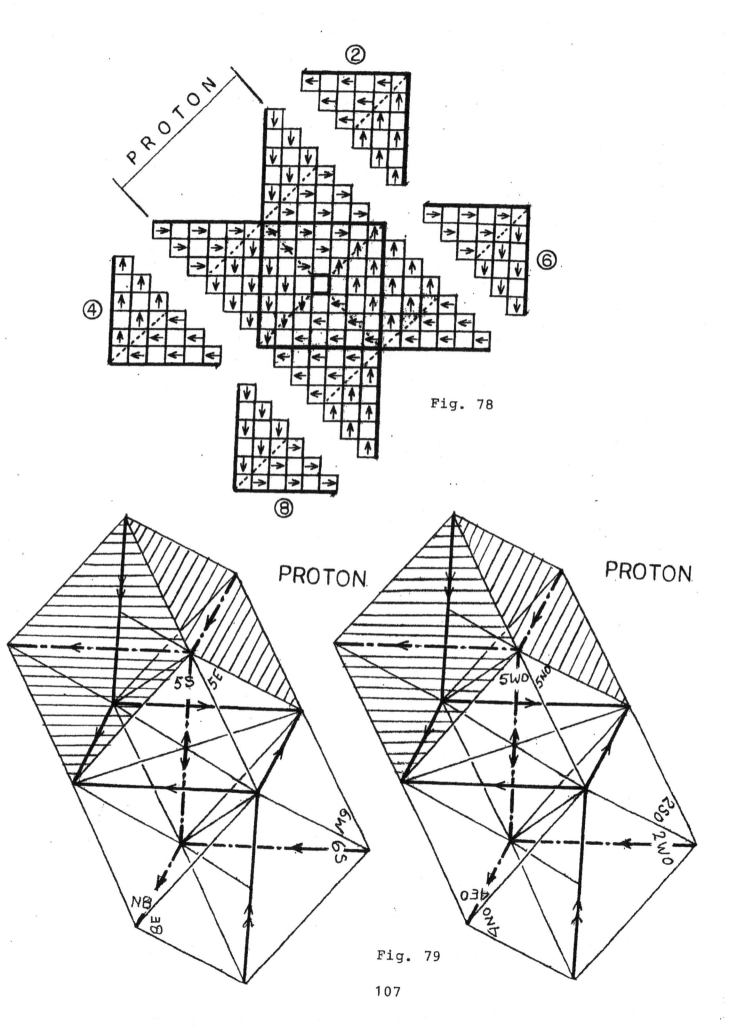

② PROTON

④

⑥

⑧

Fig. 78

PROTON

PROTON

Fig. 79

107

<u>PLEASE NOTE:</u>

The shaded scars all have at their center a dash/dot centerpost line with an arrow in it that makes it easier as how to connect with another proton, because their arrows need to line up.

<u>IN ORDER TO KEEP THE EXPLANATION EASIER TO FOLLOW,</u> the protons are still being shown with their pointed ends that they had lost to the formation of the neutrino in Fig. 43a.

The proton of Fig. 80 is shown here with a shaded diamond shaped scar of a departed quarter electron at its upper left front. Fig. 81 is a proton with a shaded scar from a departed quarter electron at its lower right backside.

It is now obvious in Fig. 82 that when we bring the two protons together in a predictable way: When we place the proton of Fig. 81 with its shaded scar structure on top of Fig. 80's proton's shaded scar, then we have a chemical bond that benefits both protons, because the scar of one proton gets covered by the inner congruent half double pyramid structure of the other proton. Both connections are now resembling one of the four neutral corners of a neutron.

The two participating proton's are now bonded in an elongated side-by-side alignment. The next page will show a crossed bonding connection between two protons.

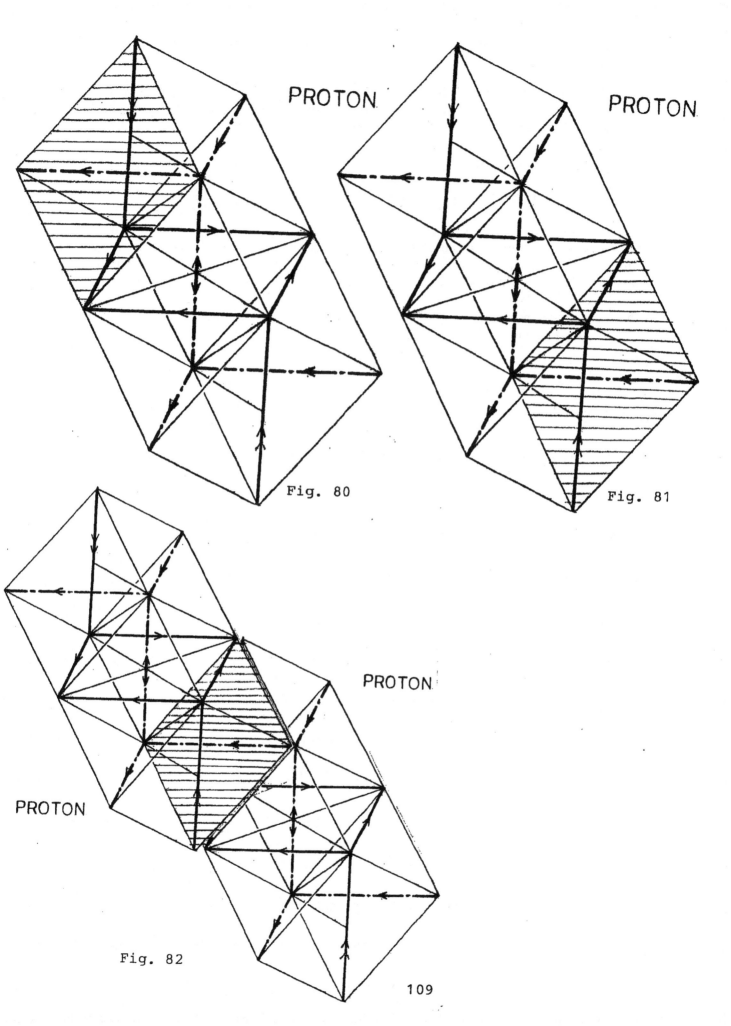

PROTON

PROTON

Fig. 80

Fig. 81

PROTON

PROTON

Fig. 82

109

Fig. 83 shows the proton of Fig. 80 here again, where we now look at the proton's upper right hand corner with its scar of its former quarter electron which is exactly congruent with the shaded scar of the internal quarter electron precursor's shape of Fig. 81, shown here again as Fig. 84.

This Fig. 84's proton's centerline is at an angle of approximately sixty degrees with the centerline of the proton of Fig. 83, which causes these two protons to bond at an angle.

When in Fig. 83, shown here again as Fig. 85, and Fig. 84 we look at both the proton's weak force and gravity respectively and when we then place Fig. 83 under Fig. 84, so that their respective arrows of their dash/dot centerpost are in alignment, then the strong two-arrow gravitational force of each proton will suck in the weaker single arrow of the other proton's weak force.

This is chemical bonding, it is somewhat like the suction hose of a vacuum cleaner hanging onto the palm of your hand.

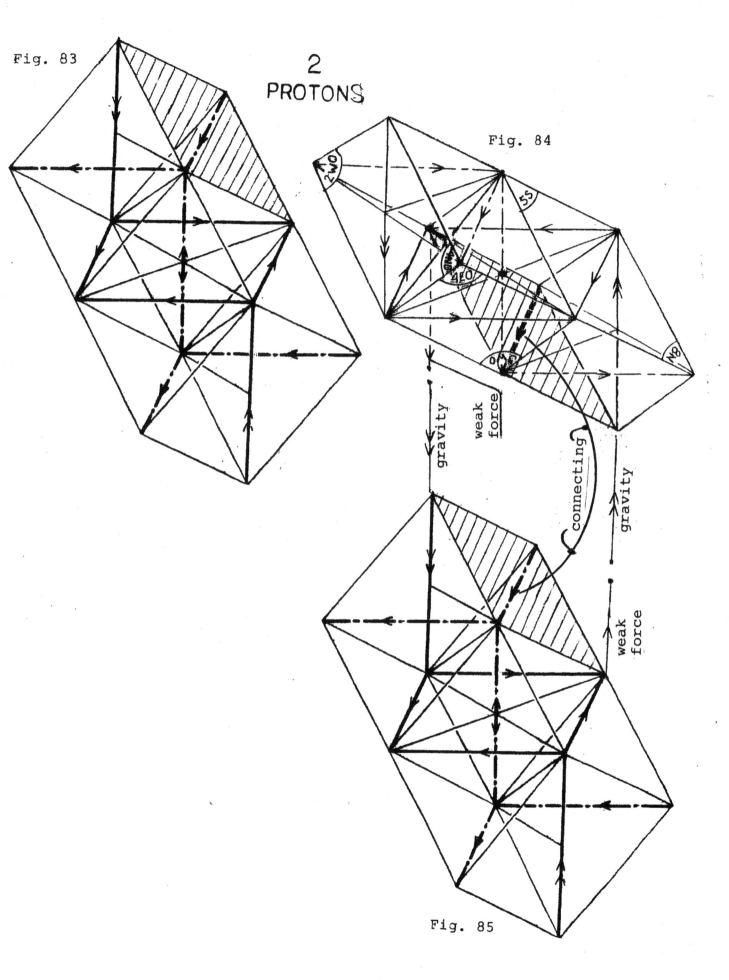

Fig. 83

2
PROTONS

Fig. 84

gravity

weak force

connecting

gravity

weak force

Fig. 85

The two protons of Fig. 85 on the previous page, still with their pointy ends which they would later lose to the neutrino, have the (shaded) aligned dash/dot diagonals of their diamond shaped scars drawn here with bold lines in Fig. 86, 87. These two diagonals at their central location will serve as a common double centerpost for these two proton's identical diamond-shaped scars.

In Fig. 86 and Fig. 87 these two protons are now shown without the pointy ends which they had lost to the neutrino, as shown in fig. 43a.

These two smaller protons are now preparing to cross-connect as in Fig. 85. Again their dash-dot diagonals will merge as shown here in Fig. 86 and Fig. 87, but their diamond-shaped scars are not flat anymore, as can be see in Fig. 49a,b,c. Each diamond scar now somewhat resembles a partly folded piece of paper.

When these two protons will bond now, then their respective scars' centerposts will be merging on top of each other, thus making a hinge, because their lost neutrino components left behind a gap between the two protons, each of them at one side of the hinge.

The above description also applies to the in-line bonding of the protons of Fig. 82.

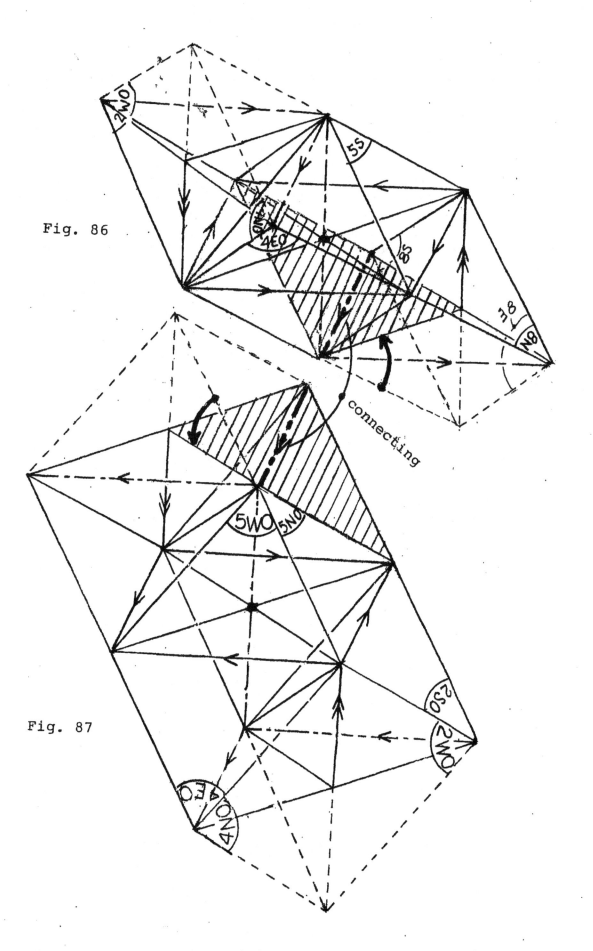

Fig. 86

Fig. 87

connecting

Returning now to the in-line bonded arrangement of the protons of Fig. 82, in Fig. 88a the hinge that is made of the two paired dash-dot arrows is exactly in line with the reader's line of sight so that we only see the hinge on end, like a point. Both protons are now shown here with some small wiggly remnants of their pointy ends which in Fig. 43a were ejected by the proton on behalf of the neutrino.

What we see now in Fig. 88a is that these two protons are only touching along their common hinge. Because there is now a gap where those lost proton ends used to be. Those vacated locations will now permit these two protons to wiggle somewhat left or right, in other words: This is a connection that is flexible, as shown by Fig. 88b and Fig. 88c. It is flexibility that permits the DNA of biology's double helix to bend and curve itself. If this were a rigid bond, then this connection would be brittle and would break, and the double helix would not be able to use its agility to make connections.

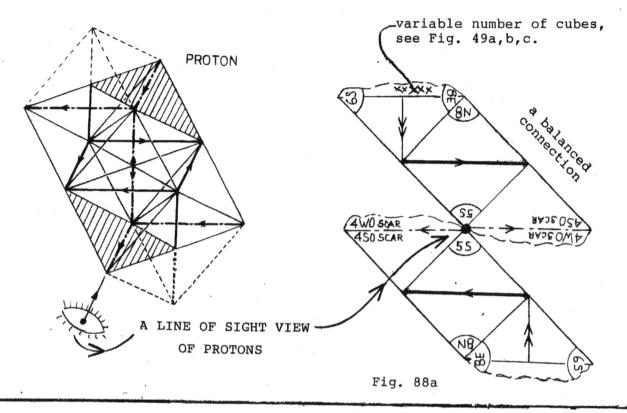

PROTON

variable number of cubes,
see Fig. 49a,b,c.

a balanced connection

A LINE OF SIGHT VIEW
OF PROTONS

Fig. 88a

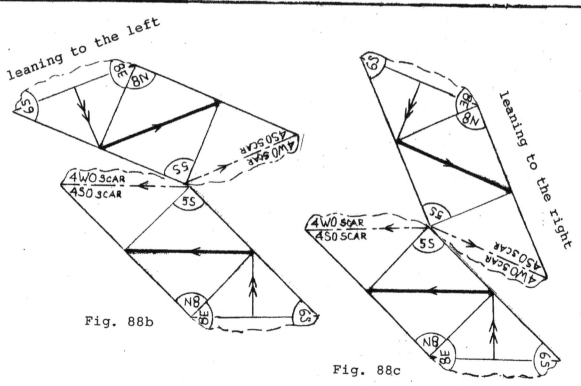

leaning to the left

Fig. 88b

leaning to the right

Fig. 88c

Helium is the first element that follows hydrogen in the Periodic Table of Elements and it consists of two protons and two neutrons which somehow have managed to fuse each other together to such a great extent that this whole group acts as if it were one homogenous entity.

It turned out to be a taxing endeavor to break its method of assembly, not having any indication of what to do with the nuclear mass of the protons and neutrons, and would all of this work out with the orbits of the electrons?

It was educational, to say the least, this theory was going in a direction that I am sure nobody had traveled before.

After eight years of trying with models of two protons and two neutrons that I had built, an attractive, justifiable solution was found that could be defended from the points of simplicity, consistency and structural reliability. The next step was then to devise a way in which this helium test case could be analyzed as to its present and future capabilities.

In the end it was necessary to devise a chart that would keep track of all the connections that the various pyramid slopes would make with each other.

Both the proton and the neutron have many triangular pyramid slopes, which in the beginning of this theory already had been given an identification code as to their exact location.

In Fig. 89, 90 only those pyramid slopes are mentioned which are actively involved in the assembly of Helium, all others are left out.

In Fig. 89 one proton and one neutron are pairing themselves into an upside down V-shape by means of the merging of their respective 6S pyramid slope with the other particle's 8E pyramid slope.

Fig. 90 is displaying exactly the same combining of the proton with the neutron, shown here upside down with respect to Fig. 89 above.

(Continued on next page)

MAKING HELIUM

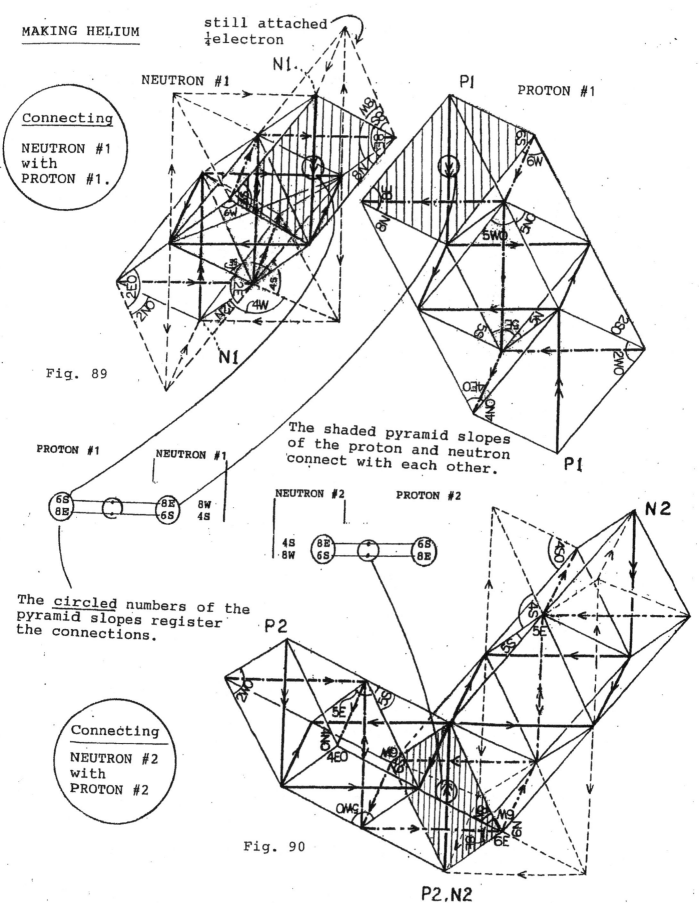

still attached ¼ electron

NEUTRON #1

N1

Connecting

NEUTRON #1
with
PROTON #1.

Fig. 89

PROTON #1

P1

N1

PROTON #1

NEUTRON #1

The shaded pyramid slopes of the proton and neutron connect with each other.

| 6S | | 8E |
| 8E | | 6S |

8W
4S

P1

NEUTRON #2   PROTON #2

| 4S | 8E | | 6S |
| 8W | 6S | | 8E |

The <u>circled</u> numbers of the pyramid slopes register the connections.

P2

N2

Connecting

NEUTRON #2
with
PROTON #2

Fig. 90

P2, N2

117

Fig. 89 and Fig. 90 are shown here again as Fig. 91a and 91b respectively, in which the double-arrowed centrally located merged baselines of pyramid slopes 6S and 8E of both V-shaped assemblies are shown in a perfect in-line alignment with each other.

When these upper and lower V-shapes are thus pushed together, then the built-in unique assembly of each proton/neutron pair will in Fig. 92 result in the packaging of all four participating central section 5's in a flat, square plane in which the two protons are side by side and the two neutrons are also side by side.

A SPECIAL NOTE:

Front he very beginning when I began to develop this theory I though that there were many options available for different ways to assembly protons and neutrons, because there were so many similar and/or identical pyramid slopes. I believed at first that in this 3-D jigsaw puzzle there would be many different paths to a solution.

That is just not so.

Each slope to slope merger can ONLY take place between two slopes which share a common baseline, like 5S with 8N, or 8E with 6S, etc., see Fig. 93.

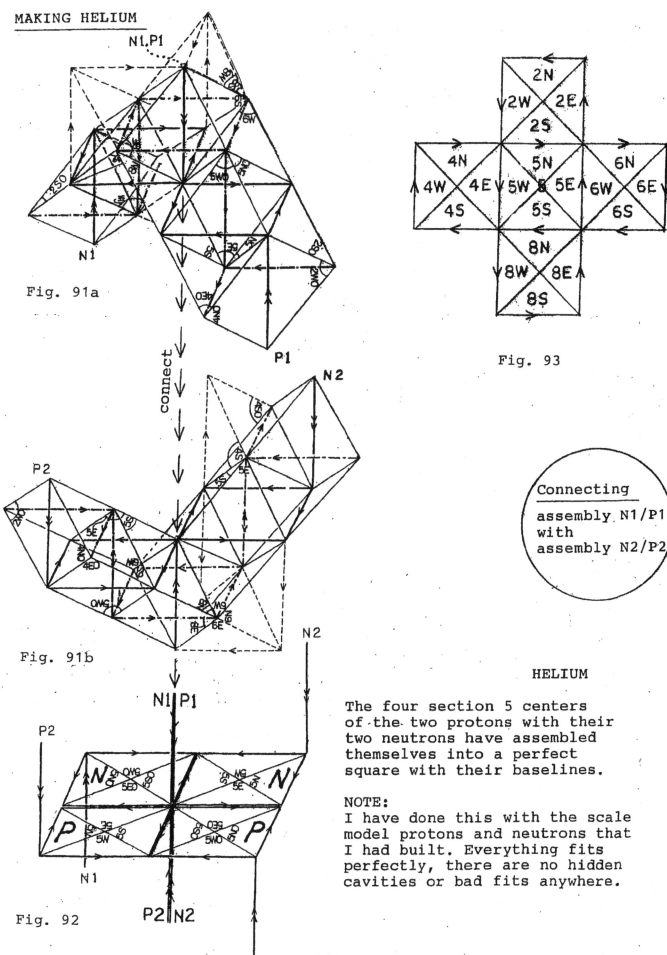

MAKING HELIUM

Fig. 91a

connect

Fig. 91b

N1 P1

P2

N1

Fig. 92

P2 N2

N2

Fig. 93

Connecting

assembly N1/P1
with
assembly N2/P2

HELIUM

The four section 5 centers
of the two protons with their
two neutrons have assembled
themselves into a perfect
square with their baselines.

NOTE:
I have done this with the scale
model protons and neutrons that
I had built. Everything fits
perfectly, there are no hidden
cavities or bad fits anywhere.

In order to give the reader insight as to how these assemblies go together an example is shown here how Helium's two neutron's slopes merge in Fig. 94 on two locations, which are 8W with 4S and 5S and 8N. These are not chemical bonds but nuclear bonds, which will be explained further on.

The key issue here is that the matrix of space, first shown here in Fig. 35, is absolutely rigid.

I have found it to be absolutely impossible to do anything other than what the matrix commands, and it is so that any arrow's direction in the matrix never can be circumvented, or ignored, or bypassed for any purpose at all.

Fig. 95 shows with double lines where the boundaries of the two neutrons are.

The connections of Helium's two protons are shown on the next two pages, with a comprehensive locater chart for all connections.

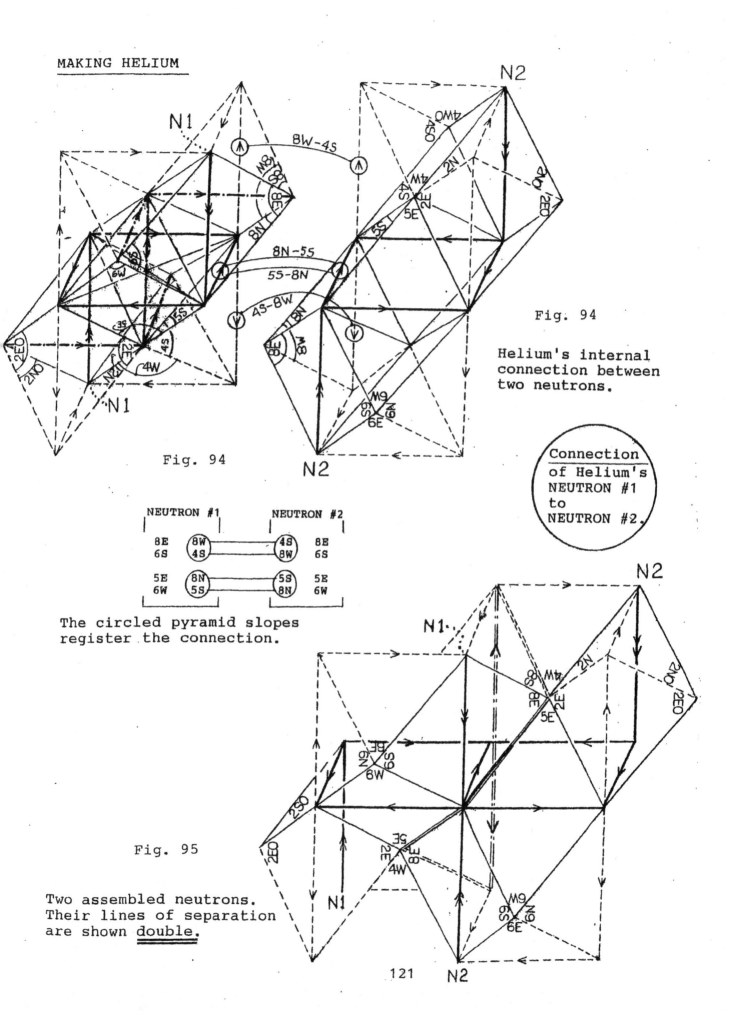

MAKING HELIUM

Fig. 94

Helium's internal
connection between
two neutrons.

Connection
of Helium's
NEUTRON #1
to
NEUTRON #2.

Fig. 94

NEUTRON #1		NEUTRON #2	
8E	8W	4S	8E
6S	4S	8W	6S
5E	8N	5S	5E
6W	5S	8N	6W

The circled pyramid slopes
register the connection.

Fig. 95

Two assembled neutrons.
Their lines of separation
are shown <u>double</u>.

At the very beginning of developing this theory it was hypothesized that the five crosswise connected double pyramids might be a possibility of nature's system that would rule the structural assembly of nature's various particles.

Identifying each pyramid slope in Fig. 5 with its own identification number was crucial in keeping track of the whereabouts of every pyramid slope, keeping in mind that every pyramid slope such as for instance 4S has below itself a matching pyramid slope 4SO, as explained earlier in Fig. 5a.

Fig. 96 gives a comprehensive oversight of how Helium's two protons and two neutrons assemble into a tight package. Fig. 96a unexpectedly became the "Rosetta Stone" of assembly patterns, which at this point I believe will keep its validity for all heavier elements beyond Helium.

HELIUM

CONNECTION CONTROL CHART

Fig. 96

PROTON #1	NEUTRON #1	NEUTRON #2	PROTON #2	
8N ———————————————— 5S				
5S ———————————————— 8N				
6W —— 5E	5S — 8N	5E —— 6W		
5E —— 6W	8N — 5S	6W —— 5E		
	8W ———— 4S			
	4S ———— 8W			
6S ———————————— 8E				
8E ———————————— 6S				
	8E ———————————— 6S			
	6S ———————————— 8E			

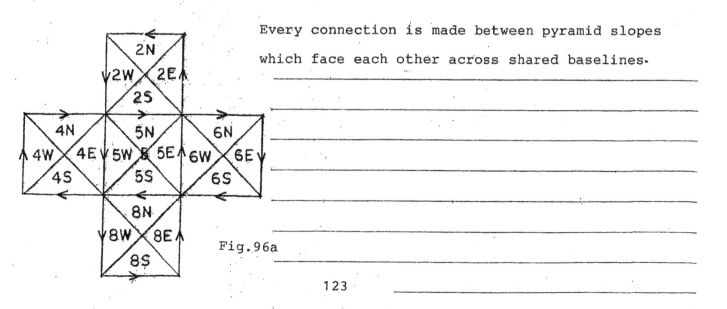

Every connection is made between pyramid slopes which face each other across shared baselines.

Fig. 96a

In the chapter on 'SPIN' it shows that gluons are classified in the same SPIN-1 category of non-gravitational force carriers such as photons and weak gauge bosons.

Having at this point reviewed various particle structures, we note that the SPIN-1 characteristic in Fig. 69c, shown here as Fig. 97, applies to the smallest constituents of weak and strong force fields which are particles called weak gauge bosons.

Fig. 55 is shows here again as Fig. 98. Per definition, quoting Brian Greene in his book 'The Elegant Universe': "The strong force is responsible for keeping quarks 'glued' together inside of protons and neutrons, and keeping protons and neutrons tightly crammed together inside atomic nuclei."

Fig. 98 shows now the SPIN-1 tetrahedron 8N/8NO of the proton is tightly 'glued' in place, and thus meets its described performance requirements.

A gluon is like a structural catalyst which enables a structure like nuclear mass to be made and exist, but not being a part of it, it is like a scaffold that is being used to build a house.

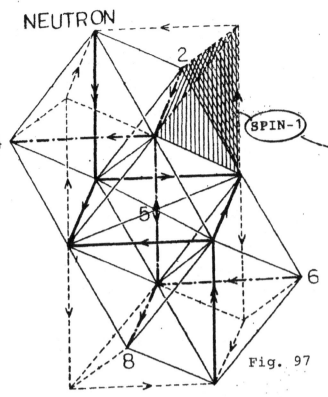

NEUTRON

2

4

5

6

8

SPIN-1

The weak force is associated
with the SPIN-1 tetrahedron
as a weak gauge boson.

Fig. 97

Spin-1...One whole tetrahedron

Non-gravitational force carriers

The photon's association with
the SPIN-1 tetrahedron can be
found in the details of the
electromagnetic wave of Fig. 65.

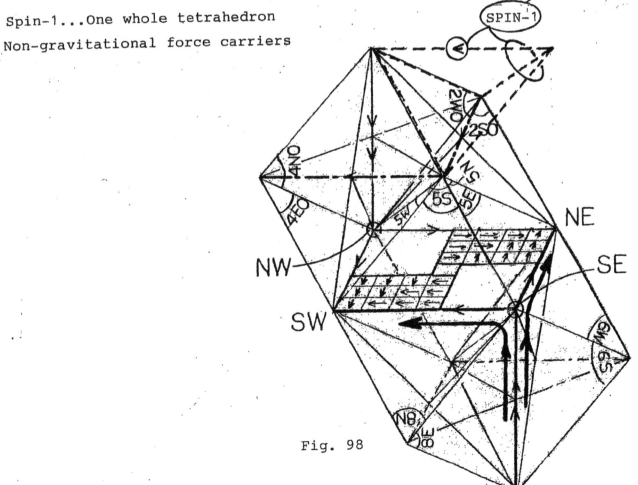

SPIN-1

NW

SW

NE

SE

Fig. 98

Every atom of the Periodic Table of Elements is known to be constructed in a manner that has the nuclear mass of its protons and neutrons clustered at its center, and that its electrons are some distance from that atom's surface.

In Fig. 92, 93, 94, 95, 96 it was shown that Helium's two protons and two neutrons are neatly packed together in a tight assembly, with no unused hidden cavities between the various surfaces of these particles.

The nuclear mass formation locations for a proton and a neutron are identical, being at the upper level of their section 5's northwest corner and at their section 5's lower and opposite southeast corner, as shown in Fig. 54.

Fig. 99 indicates the locations of the nuclear mass of the two protons and two neutrons. As a marker, a large oval indicates that the formation location of a nuclear mass unit is located above Helium's quadruple baseplate, a small oval marks a nuclear mass formation location that is below Helium's quadruple baseplate.

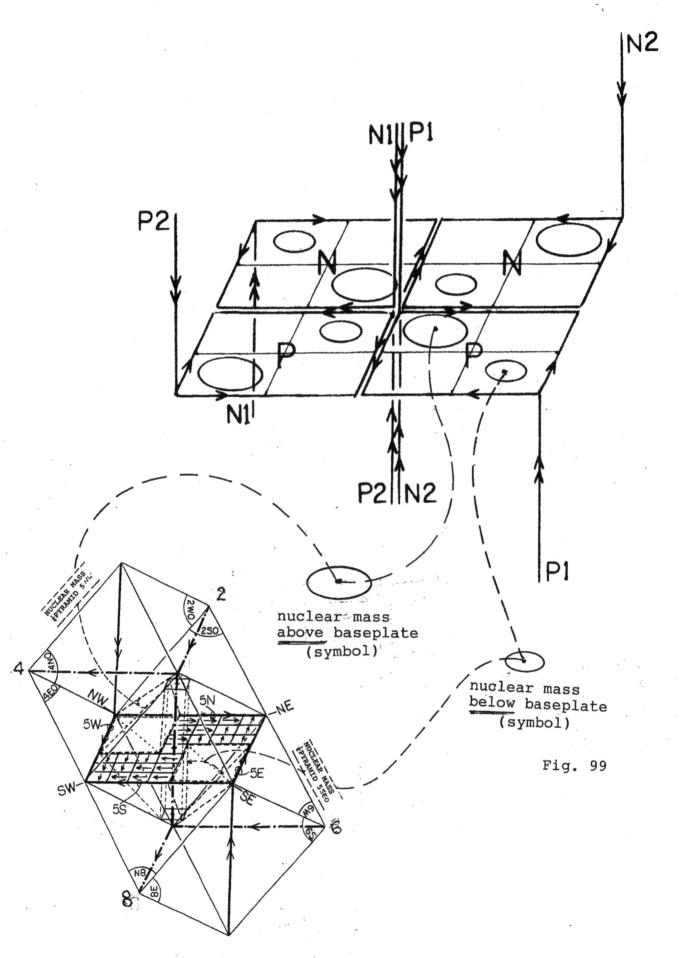

nuclear mass
above baseplate
(symbol)

nuclear mass
below baseplate
(symbol)

Fig. 99

It was shown in Fig. 52a, 52b, 54 that nuclear mass was created in neutrons and protons when the pyramid slopes of centrally located section 5's bottom-to-bottom pyramids were merged with the folded over pyramid slopes 2S, 4E, 6WO and 8NO respectively. These folded over pyramids were the quarks which produced the nuclear mass, as explained earlier.

Using the north and west portion of section 5 as an example, we see that its created nuclear mass is located <u>above</u> section 5's pyramid baseplate, in quarter pyramid 5N and quarter pyramid 5W. Also located <u>above</u> section 5's pyramid baseplate are quarter pyramid 5S and quarter pyramid 5E, which so far have been untouched.

However, when the two protons and two neutrons are being tightly packaged when Helium is being made, we see then in the locator chart shown here as Fig. 100 that in Proton #1's SE corner that

- Proton 1's Slope 5S has merged with Proton 2's Slope 8N
- Proton 1's Slope 5E has merged with neutron 2's Slope 6W.

Neutron #1 slope 8N and neutron #1 slope 6W act now as new quarks, thus making an additional nuclear mass in 5SE.

Fig. 100 indicates that Helium's tight packaging of its protons and neutrons has made new single mass units, which means that Helium somewhere else must have lost four of its original single mass units, in order to keep Helium's proper mass count of 4 the same.

How was that done?

HELIUM

NEW NUCLEAR MASS CREATION DUE TO PACKAGING

PROTON #1	NEUTRON #1	NEUTRON #2	PROTON #2	
8N —————————————————— 5S				These are new NUCLEAR MASS connections.
5S —————————————————— 8N				
New nuclear mass	New nuclear mass	New nuclear mass	New nuclear mass	
6W——5E	5S——8N	5E——6W		
5E——6W	8N——5S	6W——5E		
	8W————4S	4S————8W		These are BONDING connections
6S ——————————— 8E				
8E ——————————— 6S				
	8E——————————6S	6S——————————8E		

Every connection is made between pyramid slopes
which face each other across shared baselines.

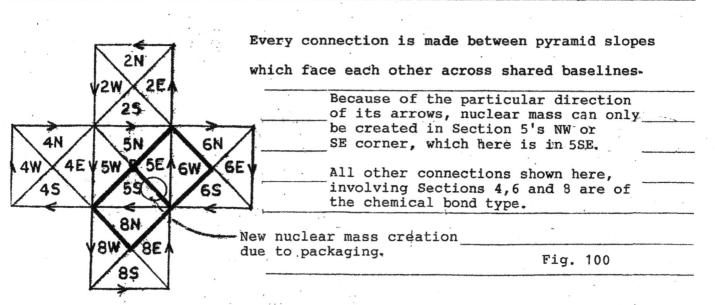

Because of the particular direction
of its arrows, nuclear mass can only
be created in Section 5's NW or
SE corner, which here is in 5SE.

All other connections shown here,
involving Sections 4,6 and 8 are of
the chemical bond type.

New nuclear mass creation
due to packaging.

Fig. 100

In order to keep track of Helium's four additional units of nuclear mass production sites, it is useful to show with four sequential drawings what is happening:

- Fig. 101a is a schematic of the placement of helium's original nuclear mass generation sites in its two protons and two neutrons.
- Fig. 101b shows only the four new 5E/5S new nuclear mass production locations, which were newly created due to the tight packaging of the two protons with the two neutrons, as shown earlier in Fig. 100.
- Fig. 101c is a composite of Fig. 101a and Fig. 101b, showing the two proton's and two neutron's original combined nuclear mass producing units as well as the four new additions.

Now something important has happened:

In Fig. 99, seen here again as Fig. 101a, we see that in Helium's center there are now from above and also from below four tightly packaged gravitational baselines moving towards Helium's exact geometric center. This awesome inward power will harden Helium's inner structure. This manifests itself in the formation of new quarks which then make new mass at four locations. The energy for doing this is drawn out of the vulnerable, structurally weaker and unsupported four remote corners of Helium itself, which cause these four corners to lose their nuclear mass. See Fig. 101d. More about that in the following pages.

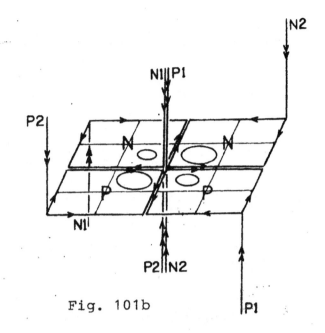

Fig. 101a

Initial assembly of helium's
2 protons and 2 neutrons.

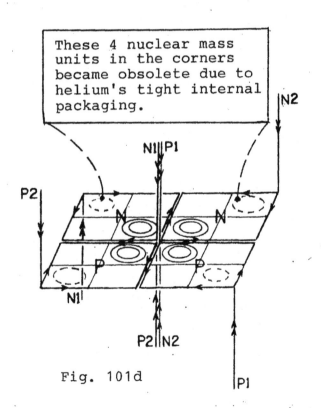

Fig. 101b

Newly created nuclear mass
due to tight packaging of
helium's protons and neutrons.

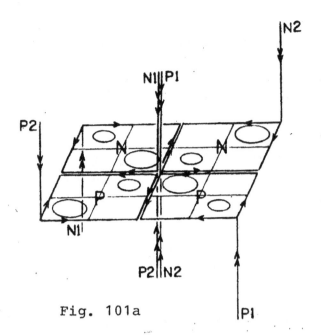

Fig. 101c

A composite of the packaging of
helium's origional and 4 newly
made nuclear mass units.

These 4 nuclear mass
units in the corners
became obsolete due to
helium's tight internal
packaging.

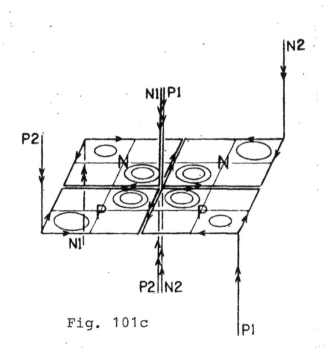

Fig. 101d

This is the final assembly of
the distribution of nuclear mass
within the confines of helium.

Fig. 102 shows in four separate drawings how the two protons and two neutrons fit into their respective place of Helium's composite square. All particle's Section 5 are in the same plane.

On Page 133: In order to keep this analysis as simple as possible, all four particles are arranged in this Helium model with their foldover section 6 and section 8 at Helium's center and with their foldover section 2 and section 4 at Helium's perimeter.

On Page 134: Fig. 103 is a fold out of Helium's inner connections when we investigate Helium's two protons and two neutrons packaged together in their rightful place. The solid black dots are on pyramid slopes which are involved in nuclear mass generation. The open rings represent merging pyramid slopes. This drawing is a pictorial presentation of Fig. 100's locator chart.

On Page 135: Fig. 104 is a follow-up of Fig. 101a,b,c,d, in which all section 2 and section 4 foldovers of Helium were removed, shown here in phantom. The result here is a greatly simplified Helium, it is clean.

On Page 136: Fig. 105 shows in four separate drawings what the remains of the two protons and two neutrons look like, they are now nearly identical. Surprise!!!

On Page 137: Fig. 106 is Helium's final version, it has perfect symmetry within itself, which makes it inert to other particles, because it needs not to 'borrow' other particle's components to become more perfect through chemical bombing. Helium will still be able to have its two electrons, because there are stationary orbit locations available in its surrounding matrix.

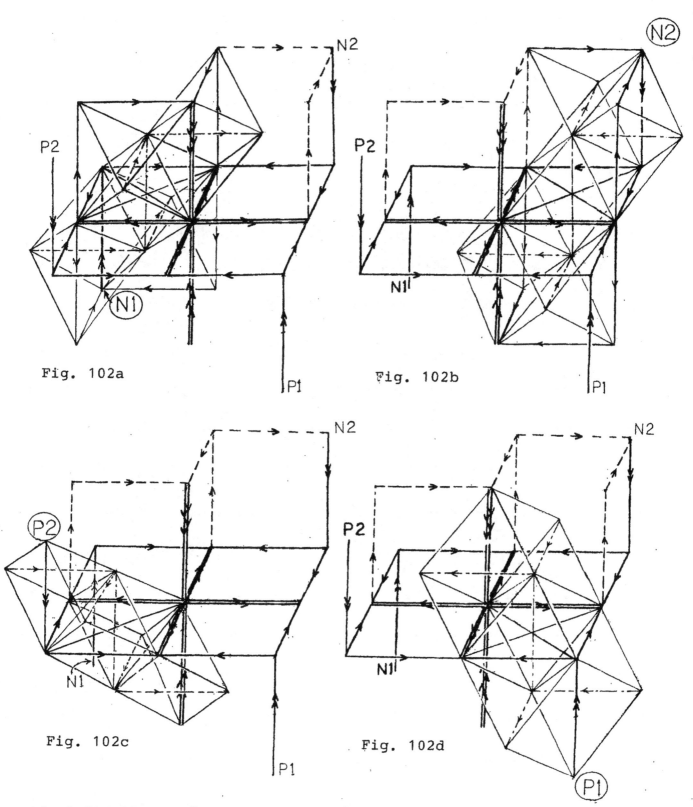

Fig. 102a

Fig. 102b

Fig. 102c

Fig. 102d

RULES OF THE MATRIX for pyramid baselines.

* A node in the matrix is like a mirror: A force arrow beyond a node
  is directionwise the mirror image of the force arrow in front of
  the mirror, no exceptions.
* Each square of each cube has opposite arrows on opposite sides.
* The sides of all cubes have arrows with CW or CCW rotation.
* The top and bottom of each cube are rotated 90° away from each other.

133

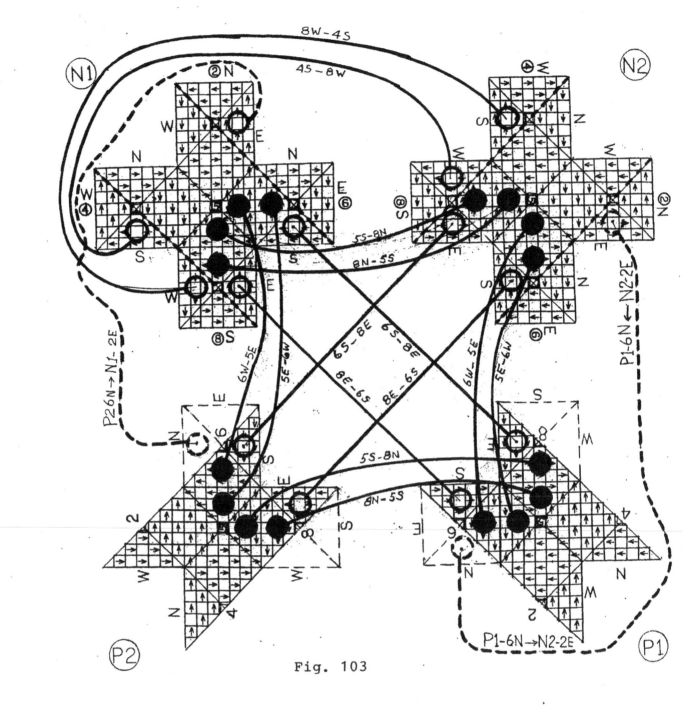

Fig. 103

This is a completely unfolded drawing of helium.
It is a pictorial rendition of HELIUM'S CONNECTION CONTROL CHART
of Fig. 100.

The connected black dots are Quark structures that are making
nuclear mass.

The connected black rings are merging pyramid slopes which are
united in a chemical bond type relationship.

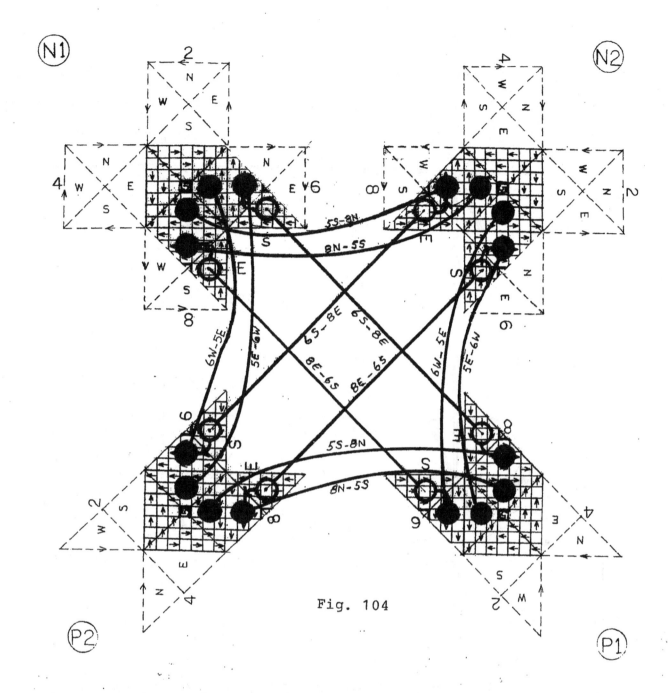

Fig. 104

It has been shown on several occasions in this theory that
structural weakness of certain particles resulted in the loss
of a portion of their structure.
Examples of that are the breakup of the neutron in Fig. 40,
the proton's loss of its pointed extremeties in Fig. 43a,
the slimming down of the neutrino in Fig. 45,
and the electron's minimal structure in Fig. 68b.

Nature is efficient. The central core of helium is held together
by a practically unbreakable lock that the quarks have put
together when they fabricated all of that nuclear mass.
Fig. 104 above shows sections 2,4 and half sections 6,8 did not
participate in the making of nuclear mass, which makes them
weak in comparison with helium's core, and they just drop away.
William of Ockham would have been pleased, had he known about this.

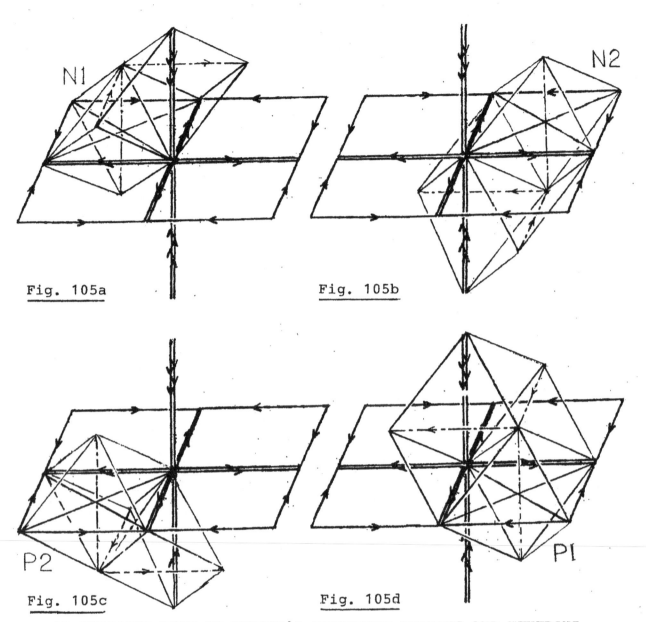

Fig. 105a

Fig. 105b

Fig. 105c

Fig. 105d

AN EXPLODED VIEW OF HELIUM'S CONDENSED PROTONS AND NEUTRONS

The loss in Fig. 104 0f sections 2 and 4 by the two protons and
the two neutrons resulted in a composite helium structure that now
consists of condensed versions of its protons and neutrons.

None of their individual components project any more beyond the
square perimeter of the composite square that will make helium.
Each of these four particles is   made by using only the input
of sections 5,6 and 8.

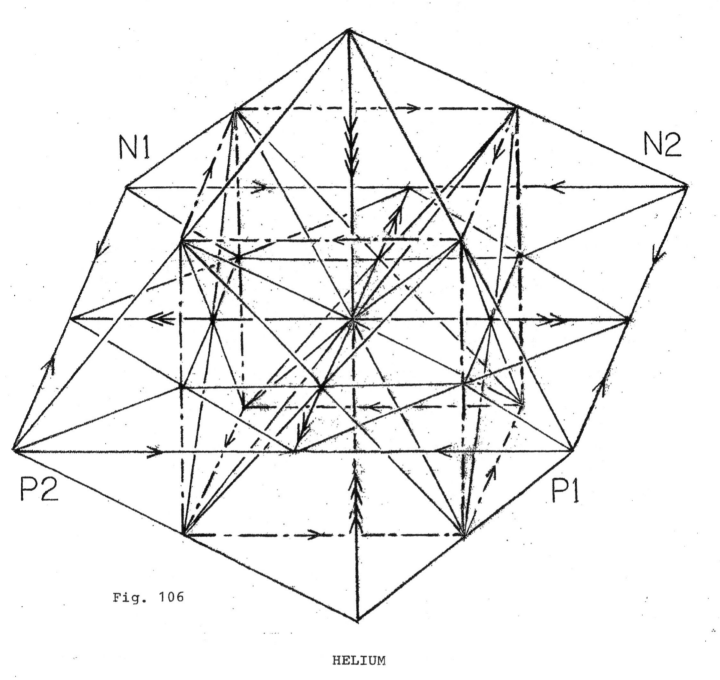

Fig. 106

HELIUM

19 Dec. 2006

Fig. 106's assembly of helium's condensed protons and neutrons
of Fig. 105abcd resulted in a aestetic, clean design that above
all is complex and simple.

Inparticular it is a revelation that helium's bottom to bottom
pyramid pair shelters a large cube inside itself.
This cube is made of twelve dash-dot vacant pyramid centerposts
which unifies the two protons and two neutrons into a single unit.

Helium's packaging of its nuclear mass in Fig. 99, 100 and 101 efficiently used those opposite corners of its protons and neutrons that were not yet occupied by their own original nuclear mass.

As an example shown here in a detailed sequence Proton P1 at Location 'A' has its own two mass cubes arranged. Location 'B' shows Proton P1 situated in Helium, which on its opposite corner has Neutron N1.

A large oval is the symbol for a nuclear mass cube that is located above the base plate, the small oval is the symbol for a nuclear mass cube below the base plate.

Due to Helium's tight packaging of its protons and neutrons, these particles are thereby induced to generate nuclear mass in those parts of their bodies that so far had not yet had an intimate contact with other particles. These close contacts between 2 individual parcels create new nuclear mass in places that are exactly opposite the original nuclear mass, resulting in a nuclear mass crossover that results in that the existing cube above Helium's base plate of Neutron N2 gets a cube below its base plate, see 'C.'

This new cube is actually located directly across Helium's own exact center, opposite the upper cube of Proton P1, and because of being at that exact location, it becomes the Newton's 3rd Law partner of Proton P1's upper cube, see 'D.' This means that Proton P1's original lower cube, located at the outside of the Helium perimeter, lost its upper level cube as a partner, and this abandoned lower cube had to disappear. Somebody had stolen his girl.

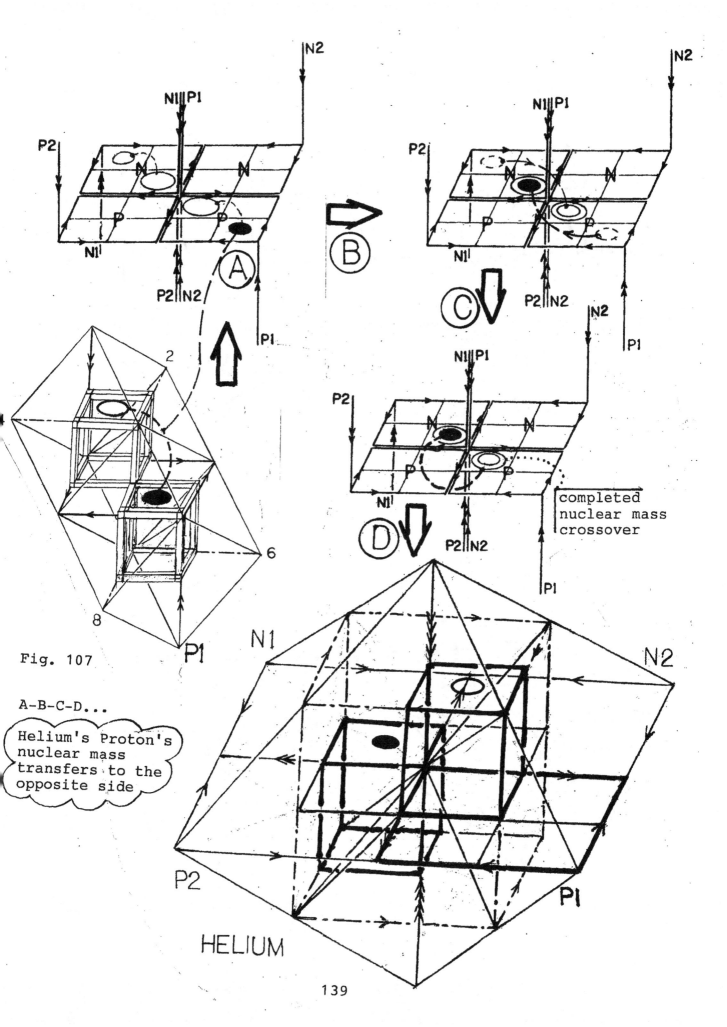

N2

N1 P1

P2

N

N

P2 N2

A

B

P1

2

6

8

P1

Fig. 107

A-B-C-D...

Helium's Proton's
nuclear mass
transfers to the
opposite side

N2

N1 P1

P2

N

N

P2 N2

C

P1

N2

N1 P1

P2

N

N

completed
nuclear mass
crossover

P2 N2

D

P1

N1

N2

P2

P1

HELIUM

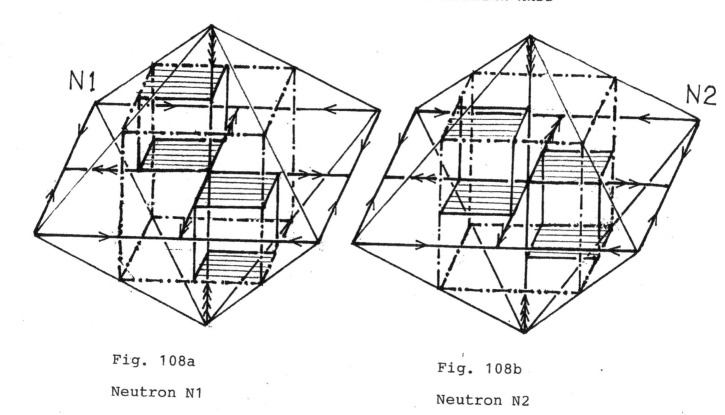

Fig. 108a

Neutron N1

Fig. 108b

Neutron N2

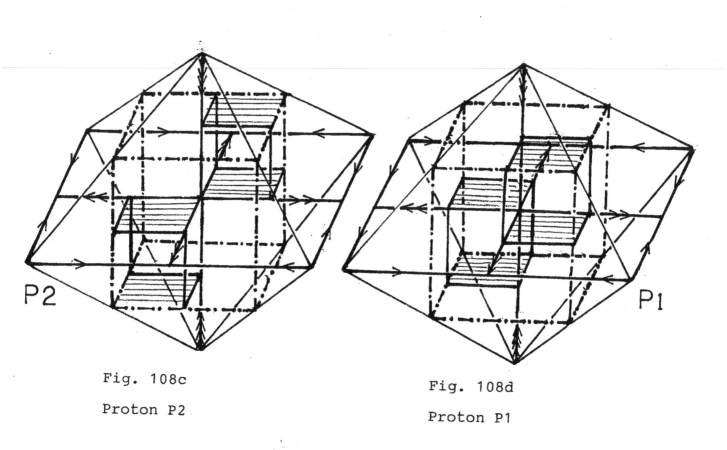

Fig. 108c

Proton P2

Fig. 108d

Proton P1

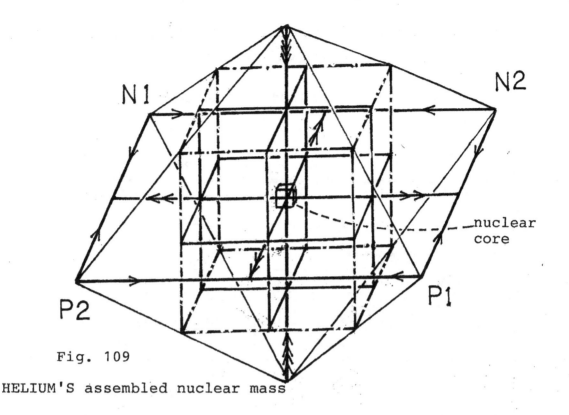

Fig. 109

HELIUM'S assembled nuclear mass

After having shown on the previous pages the method of repackaging of
the nuclear mass pairs of the protons and neutrons in helium, it can now
be visualized in helium by using the drawings of Fig. 108a,b,c,d.

What we end up with in helium's figure of 109 is that helium has now
8 cubes into one large cube.
In Fig. 54a is shown how the proton's 2 cubic mass units share a single
cube between them in the proton's center.
In helium's case, in order to store the 8 mass cubes of its 2 protons and
its 2 neutrons, it breaks up that one-cube overlap and stores the 8 mass
units in a 2x2x2 cube, that has a new inner core that contains 4 times
the volume of of the origional single cube proton/neutron overlap.
Because of its location in helium's exact center, this mutual mass core
is a small part of each of its surrounding 8 large mass generating cubes.

Going back now to the Chapter 'Homing in on nuclear mass', the pairing of
a proton with a neutron created a combined mass number of 3674.0000,
which is an average of of 1837.0000 for each the proton and the neutron.

The interesting thing about this is that the neutron's mass is
1838.2656 - 1837.0000 = 1.2656 <u>more</u> than that average, and the proton is
1837.0000 - 1835.7344 = 1.2656 <u>less</u> than that average, exactly the same !!

Is there something special about that 1.2656 number?
Its square root is $\sqrt{1.2656} = 1.124988$, which is practically 1.125,

which is $\dfrac{9}{8} = \dfrac{3^2}{2^3}$.

The simplicity of this fraction reminds one of neutron's $(3\tfrac{1}{2})^6$ nuclear mass.

DEUTERIUM

CONNECTION CONTROL CHART

PROTON #1	NEUTRON #1	NEUTRON #2	PROTON #2	
8N ———————			——— 5S	
5S ———————			——— 8N	
6W —— 5E	5S — 8N	5E —— 6W		
5E —— 6W	8N — 5S	6W —— 5E		
	8W ——— 4S			
	4S ——— 8W			
6S ———————		——— 8E		
8E ———————		——— 6S		
	8E ———————		——— 6S	
	6S ———————		——— 8E	

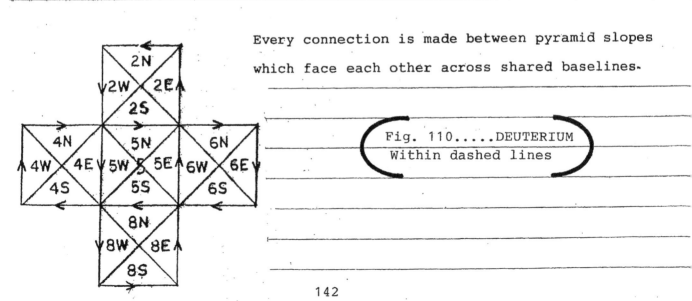

Every connection is made between pyramid slopes which face each other across shared baselines.

Fig. 110.....DEUTERIUM
Within dashed lines

142

TRITIUM

CONNECTION CONTROL CHART

PROTON #1	NEUTRON #1	NEUTRON #2	PROTON #2	
8N ———————			——— 5S	
5S ———————			——— 8N	
6W —— 5E	5S — 8N	5E ——— 6W		
5E —— 6W	8N — 5S	6W ——— 5E		
	8W ——————— 4S			
	4S ——————— 8W			
6S ———————		——— 8E		
8E ———————		——— 6S		
	8E ———————	——— 6S		
	6S ———————	——— 8E		

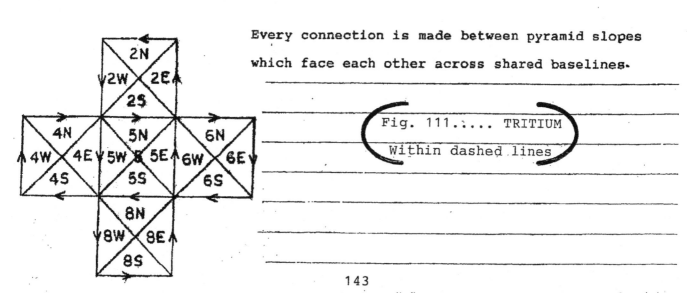

Every connection is made between pyramid slopes which face each other across shared baselines.

Fig. 111. .... TRITIUM
Within dashed lines

Throughout the development of this theory there were major problems with finding the path of an electron's orbit around a proton.

I know that all of us have been very influenced by seeing the moon in its steady monthly orbit around the earth, as well as by the yearly, steady and predictable curved orbit of the earth around the sun.

Maybe as a consequence thereof, scientists have usually pictured an electron's orbit around a proton as a smooth, elliptical path that is traveled at a blinding speed.

Experiments have determined that an electron is orbiting at a higher or lower level around a proton being subject to the variable content of an electron's energy as it relates to its proton's energy.

In this theory an electron's orbit is not continual, but intermittent. For example, in Fig. 112 of the proton:

- A – The perception of an electron's high orbital speed is actually the observation of a stationary, intrinsic velocity of the arrow's forces within the electron.
- B – An electron is perceived to have a higher orbit when it has a stationary location that is farther away from its proton.

ORBITING HELIUM

Fig. 112

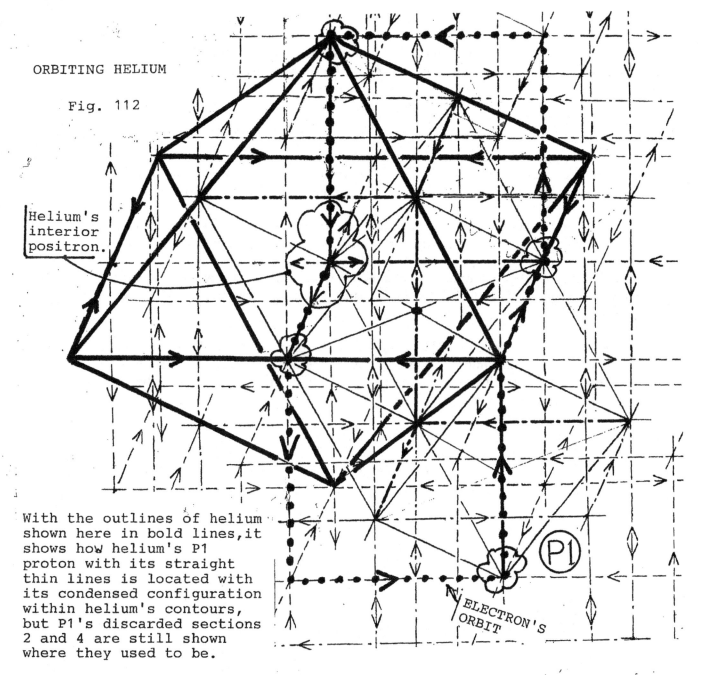

Helium's interior positron.

With the outlines of helium
shown here in bold lines, it
shows how helium's P1
proton with its straight
thin lines is located with
its condensed configuration
within helium's contours,
but P1's discarded sections
2 and 4 are still shown
where they used to be.

This is important to know, because this proton's electron will use
those section's 2 and 4 patterns as a part of its circuit around
and through the proton.
At the upper right of this proton is also a path for the electron
through the matrix where one of the neutron's now discarded quarter
electrons used to be.
The orbit's corners that are marked with a small cloud have the same
arrow pattern as an electron. These locations provide for a
'reststop' for an electron, where it may stay for a while or move on.

The arrow pattern of helium's central core, shown here with a larger
cloud around it, matches exactly the arrow pattern of the positron
of Fig. 42c.
When the electron arrives there, either from straight above or below,
it then annihilates into the large positron, after which it can
start anew by going either N,E,S or W, wherever it is drawn to.
A second electron travels a similar path, in this way both electrons
provide their import/export energy services for helium.

145

For clarity proton P2 electron's orbit is shown here by itself in Fig. 113 so that we can see that it is upside down as it relates to the orbit of proton P1, otherwise they are identical, they follow an identical path to complete their full orbit.

Describing Fig. 114.

In physics it is a well known fact that two electrons cannot occupy the same orbit. Here, in Helium's case, we see in Fig. 114 that the two electron circuits share a path that is at the shared internal border of protons P1 and P2, which cannot be seen by observers from the outside.

Each electron circuit of an electron moves along the arrow's direction of eight pyramid baselines. Two of those baselines belong to structural components of the proton's section 5, and one of those baselines is a half length of Helium's main centerpost. The other five baselines belong to the proton's external mantle of its non-solid matrix, which consists of a two-baseline and a three-baseline section, in which the electron has a chance to become "airborne," so there is some sort of orbit after all!!

In the shared baselines of P1 and P2's common border protons P1 and P2 and their respective electrons have an opportunity to share and/or exchange energy, so that energy can be delivered or taken away from all other locations within the structures of their own combined bodies as well as from their surrounding matrix.

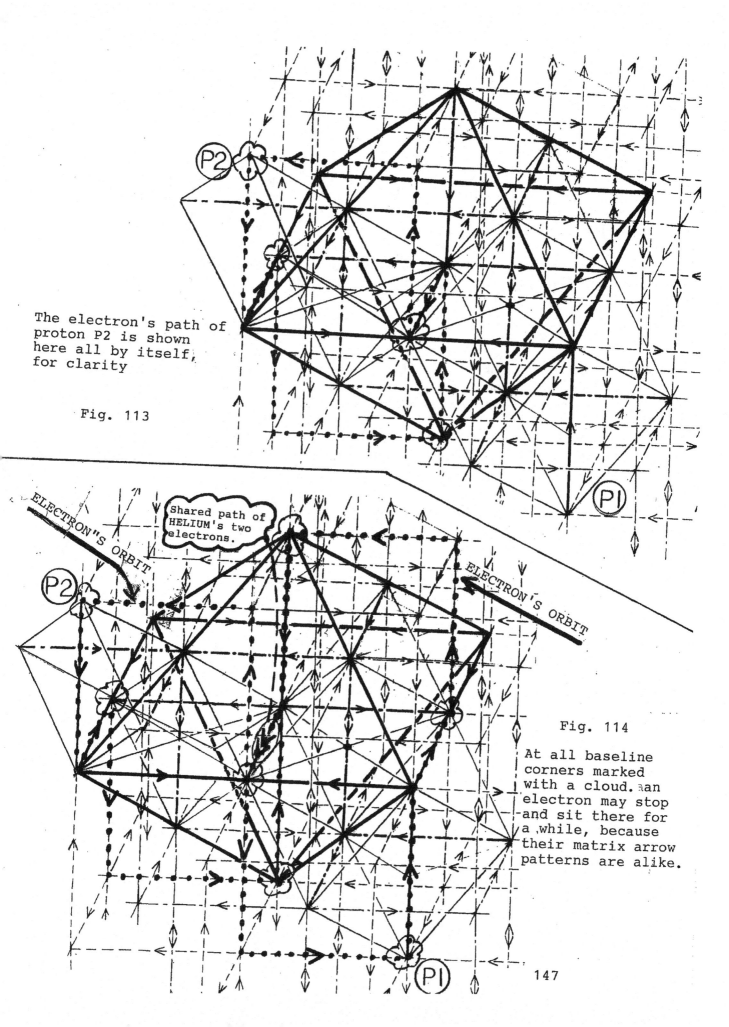

The electron's path of proton P2 is shown here all by itself, for clarity

Fig. 113

Shared path of HELIUM's two electrons.

ELECTRON'S ORBIT

ELECTRON'S ORBIT

P2

P1

Fig. 114

At all baseline corners marked with a cloud, an electron may stop and sit there for a while, because their matrix arrow patterns are alike.

147

Fig. 115.

Having found by now that Helium's overall configuration in this theory consists of eight bottom-to-bottom pyramids which have their square pyramid baseplates connected along their proton and neutron baselines, we now also know that Helium's pyramid slopes make a 45 degree angle with Helium's pyramid baseplate.

It is obvious now that Helium's external configuration is on a linear scale twice as large as any of the neutrons original sections 2, 4, 5, 6 or 8, but they all do share that 45° slope feature of their pyramids.

This takes us to the next step in Fig. 116:

Alpha radiation consists of a stream of Helium ions, which strongly suggests that those elements which emit that radiation consist of a tightly packed group of Helium atoms, which would follow an assembly pattern as used for the neutron, as shown in Fig. 28, 29.

Building on that thought: Neon 20 with its 10 protons and 10 neutrons would then be made of five Heliums, as shown here, their assembly most likely might mimic the sequence that made the neutron.

Lithium would consist of two Heliums, with one of the protons missing.

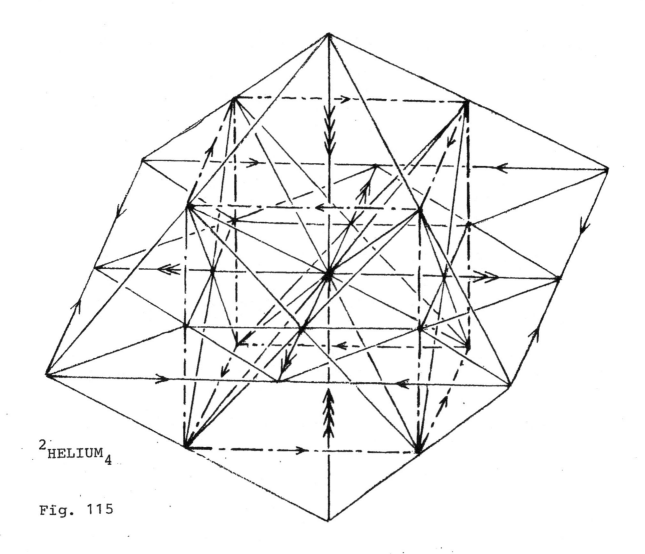

$^{2}HELIUM_{4}$

Fig. 115

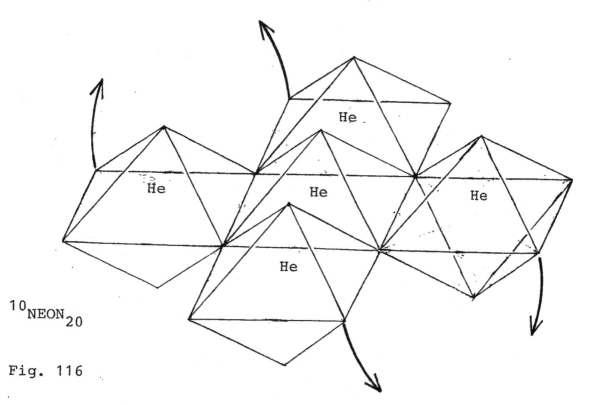

$^{10}NEON_{20}$

Fig. 116

The characteristics of the proton were reviewed here in these pages with its Fig. 66a, shown here as Fig. 117, as a reference for analyzing Helium's structure.

In Fig. 118 Helium's site of electromagnetic radiation is found at each end of the four centerposts of its two protons and neutrons.

Helium's gravitational pull is concentrated at the top and bottom apex of its central 'vertical' axis which at both locations consist of four bundled baselines.

The attracting nature of these two sites of gravitational pull is counterbalanced by the repelling activities of the eight weak forces which are located as four pairs of opposing forces in the exact middle of Helium's straight sides.

Helium is inert, because: It has superior fulfilled symmetry throughout, therefore it does not need any kind of fulfillment in any part of its structure, which means it will not connect with anybody.

And last but not least: This whole Helium package by itself may perform as alpha radiation that may be emitted from heavier elements.

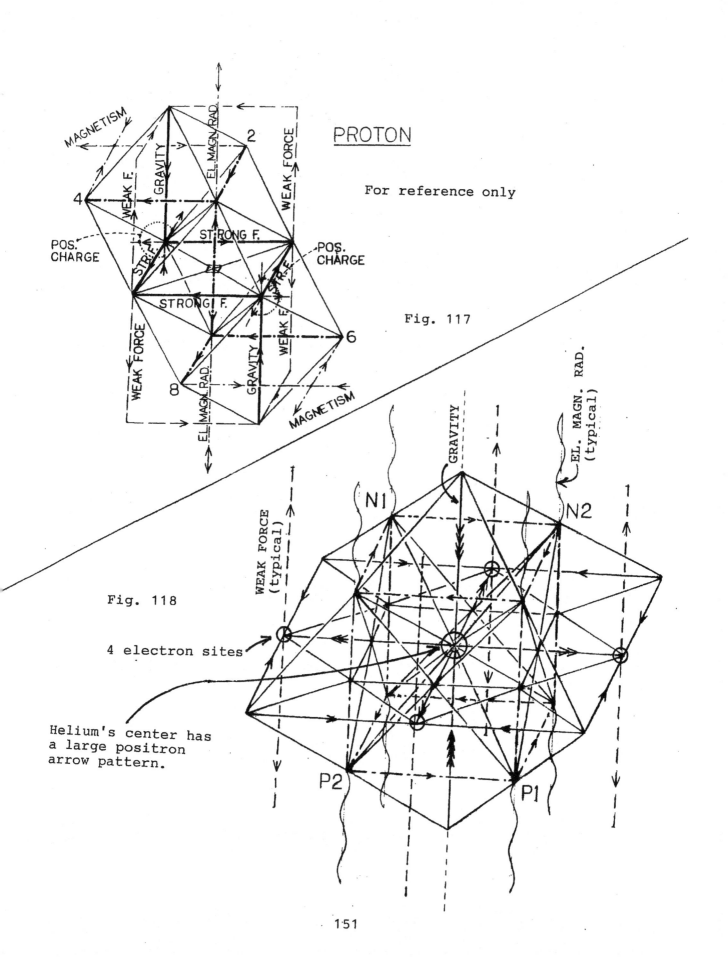

PROTON

For reference only

Fig. 117

Fig. 118

4 electron sites

Helium's center has
a large positron
arrow pattern.

151

We can take another look now at Helium's individual pyramids.

Starting with Helium's big square, one easily recognizes its four bottom-to-bottom pyramids in Fig. 119b, and Fig. 119a and Fig. 119c above and below are alike and they both will be able to provide a snugly fitting upper apex and lower apex for Helium's own larger composite pyramid.

This pointy top and its congruent pointy bottom are both exactly identical to the positron of Fig. 42c, which also is made of four half double pyramids which are assembled in a manner that represents two interlocked INSIDE OUT pyramids.

Helium thus consists of four positive inside out double pyramids that are sandwiching four negative standard double pyramids.

Helium's interplay between standard and inside out pyramids is (almost) perfect, it is an effort for nature to return to the womb, because if Helium's assembly were 100% perfect it would disappear in a small flash, just like an electron and a positron, when they meet.

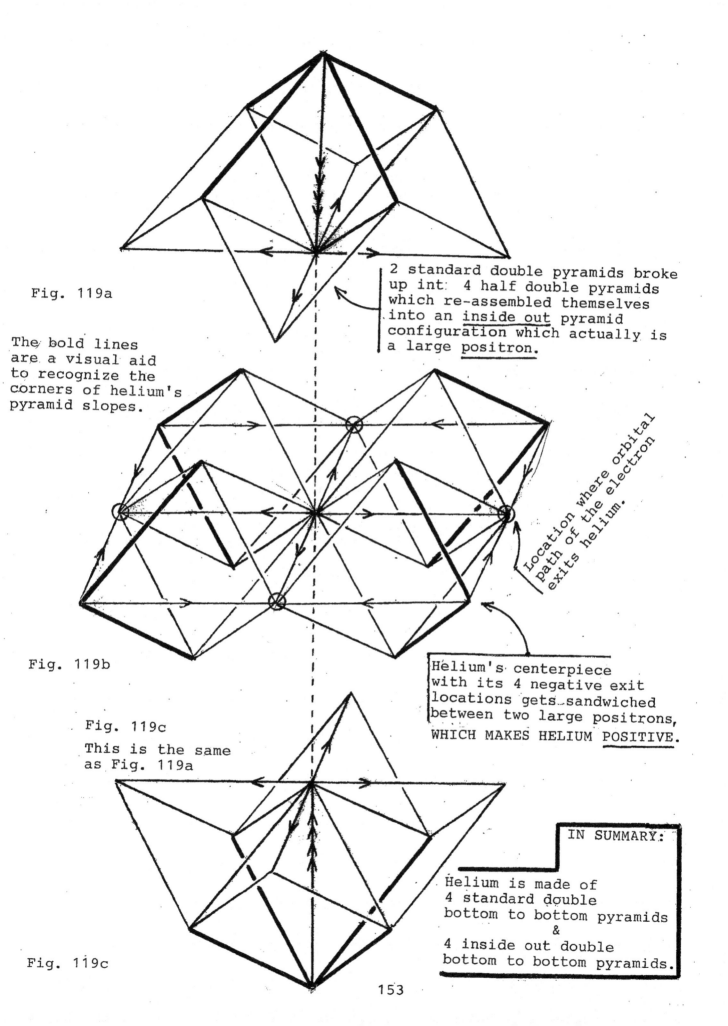

Fig. 119a

The bold lines are a visual aid to recognize the corners of helium's pyramid slopes.

2 standard double pyramids broke up int. 4 half double pyramids which re-assembled themselves into an inside out pyramid configuration which actually is a large positron.

Fig. 119b

Location where orbital path of the electron exits helium.

Helium's centerpiece with its 4 negative exit locations gets sandwiched between two large positrons, WHICH MAKES HELIUM POSITIVE.

Fig. 119c
This is the same as Fig. 119a

IN SUMMARY:

Helium is made of 4 standard double bottom to bottom pyramids & 4 inside out double bottom to bottom pyramids.

Fig. 119c

153

In the past century several proposals have been presented as to the number of dimensions we have to deal with in the Studies of Physics.

So far there seems to be no consensus as to the number of dimensions, but generally there are some trends that favor the establishment of the passing of 'Time' as the 4th dimension, even though this hardly improved our understanding of it at all.

In this theory, "Decoding the Periodic Table," the passing of time was never considered as being a candidate that could serve more or less in a manner that would support the conventional 3-D coordinate system.

Here is why:

- Forces are real, they can move in any direction within the concept of the 3-D coordinate systems of Fig. 120.

- A space can be filled with a force, but it can also not have a force at all. Vacant, unfilled space is as important as filled spaces, it provides separation and opportunity for all physical events to take place.

- Time is not a force, it is not open space of filled space or filled space. Time is a concept of mankind: Because we have a brain, we have a memory. The lapse between two events we call time, such as sunrise/sunset. Nature has no brain, no memory, there is no way that time can PHYSICALLY INTERACT with forces or distances, and that being the case, the passing of time under no circumstances can be an equal partner of any of the three force dimensions.

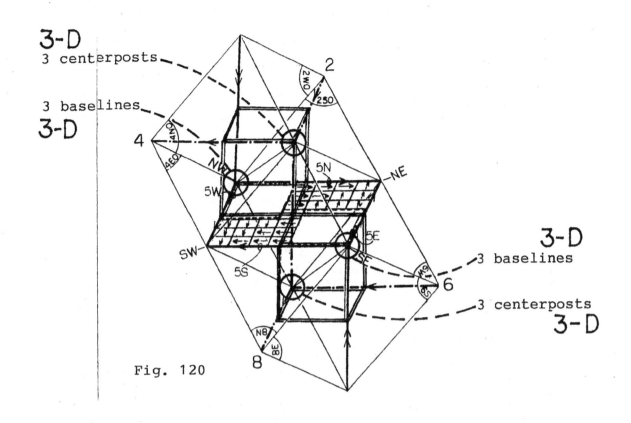

3-D
3 centerposts

3 baselines
3-D

Fig. 120

3-D
3 baselines

3-D
3 centerposts
3-D

Fig. 120 shows its neutron's section 5 with the circled origin of a
baseline 3-D coordinate system on its NW corner and also one on its
SE corner. Likewise, the upper and lower end of this proton's
centerpost also have a 3-D coordinate system, but these are for the
inside-out features of the proton.

Note that both of the two cubic mass volumes are locked in by a
baseline 3-D unit and a centerpost 3-D which are at opposite ends
of those mass cubes.
This means that all of the structural components of each baseline's
3-D system are opposed by identical structural components of the
centerpost system, as is required by Newton's 3rd Law of opposing
forces.

All of the above means that the baseline's 3-D coordinate system
can not exist without the centerpost's 3-D coordinate system,
and therefor overall this is a 6-D coordinate system.

The following is a continuation of the Chapter: "Homing in on Nuclear Mass': The formula $(3\frac{1}{2})^6$ mass posed the question of: $3\frac{1}{2}$ of WHAT? That "$\frac{1}{2}$" really drove home the notion that this "$\frac{1}{2}$" was one half of something that readily can be split into 2 equal halves. Newton's 3rd Law came to the rescue:

A 'package' of [a force with its reactive force] is activity wise a zero to the outside world. When the foothold of a reactive force is removed, then the potential of the presence of the original force gets lost.

The number "$3\frac{1}{2}$" therefore consists of 3 fully neutralized pairs, plus half a unit, which is a single unrestrained force. When that half unit of unrestrained force lost its opponent, then the entire row of the 3 packaged pairs of forces lost their opponents because all forces get pushed over like a row of falling dominos.

Fig. 121 shows one of its cubes with its $3\frac{1}{2}$ Newton force pairs in this proton's Section 5, of which its NW and SE corners are shown vacated. This means that each of those paired force units that originally had a zero force balance at the time of its creation as shown earlier in Fig. 4, 5 and 5a, will now due to their fold over become all unbalanced and forceful in their collision with opposing forces which results in the formation of nuclear mass.

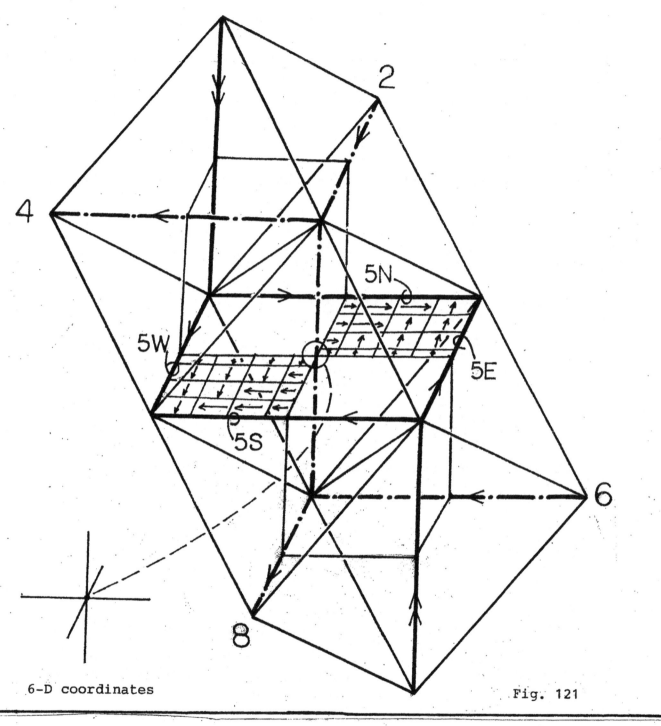

4

2

5N

5W

5E

5S

6

8

6-D coordinates

Fig. 121

$(3\frac{1}{2}\times3\frac{1}{2}\times3\frac{1}{2})$ x $(3\frac{1}{2}\times3\frac{1}{2}\times3\frac{1}{2})$ = 42.875 x 42.875 = 1838.2656 = $(3\frac{1}{2})^6$.
volume   x   volume   = volume x volume = nuclear mass

$3\frac{1}{2}$	12.25	42.875
Force	Area	Volume
	forces	forces

This applies to that single
small cube that the two
overlapping corners share at the
center of the proton/neutron.

Measured		Converted	
Electron	0.511 MeV	Electron	1.0000
Neutron	939.5731 MeV	Neutron	1838.6949
Proton	938.2792 MeV	Proton	1836.1636

Nov. 16, 2008 calculation: New Neutron Mass = $3\frac{1}{2}^6$ = 1838.2656.

Adjusting Proton Mass with reference to new Neutron Mass:
$\underline{\text{New Neutron Mass 1838.2656}}$ x Old Proton 1836.1636 = New Proton = 1835.7348 (A)
Old Neutron Mass   1838.6949

Adjusting the electron accordingly:
$\underline{\text{Measured Neutron - 939.5731}}$ = 0.5111193 MeV (new electron)
New Calc. Neutron – 1838.2656

Recalibrating NEW Proton Mass versus new Electron .5111193 MeV
$\underline{\text{Measured Proton – 938.2792 MeV}}$ = 1835.7342 ("B")
New Electron Calc. 5111193 MeV.

Reviewing Neutron/Proton combining in 2 ways:

#1. New Neutron - 1838.2656
New Proton "A"   - $\underline{1835.7348}$
            Total   3674.0004

#2. New Neutron - 1838.2656
New Proton "B" - $\underline{1835.7342}$
            Total   3673.9998

Calculation #1 and #2 are at opposite sides of 3674.0000, which is what nature most likely would want, so that heliums mass would be a perfect 7348.0000.

If the measured proton were 938.2793, which is only 0.0001 more than its measured 938.2792, which would be an acceptable tolerance, then the new proton's mass would be
$\underline{938.2793}$ = 1835.7344. (C)
.5111193

A neutron/proton combination would then be:
New Neutron 1838.2656 + New Proton "C" – 1835.7344 = 3674.000, which is perfect.

Helium, with its 2 neutron/proton combinations then has a total nuclear mass of 2 x 3674.000 = 7348.0000, which would then represent an atomic weight of 4.00000, if the unit of atomic weight measure would be chosen as the new standard of 1/4 of 7348.0000 = 1837.0000 (nuclear mass).

In this theory, helium is the building block that is capable of constructing all elements of the Periodic Table.

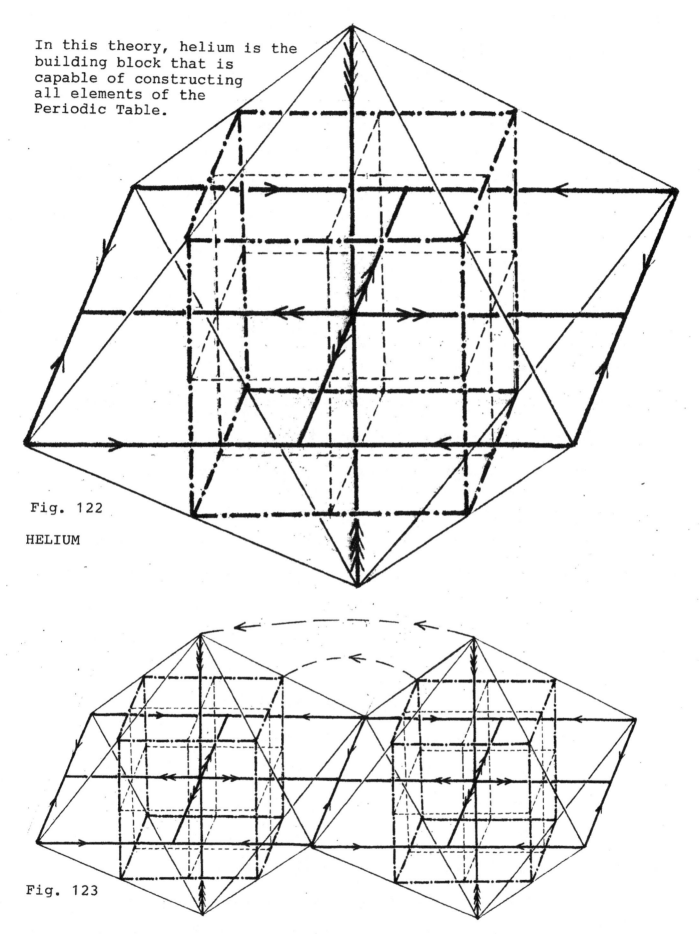

Fig. 122

HELIUM

Fig. 123

Helium to helium connections are made at a right angle fold over.

The beginning pages of this theory describe the formation of pyramids in Fig. 1, 2, 3, 4, 5 and 5a.

After an extensive analysis of that structure and also of a variety of related structures, we see then how in earlier Fig. 5a, consisting of 5 double pyramids, they are connected in the shape of a 'plus' sign, as in Fig. 28.

Folding their double pyramids 2 and 4 up till their pyramid slopes 2S and 4E come to rest respectively on double pyramid 4's Slope 5N and Slope 5W, a tighter structure has been created.

In a similar manner, double pyramids 5 and 8 are folding down till their pyramid slopes WO and 8NO respectively come to rest on section 5's double pyramid slopes 5EO and 5SO.

Upon completion of that folding scenario, a neutron had been created, Fig. 29.

Of particular note in Fig. 124, 125 is that 5 Heliums can perfectly imitate the formation method of the neutron. It is striking that each large centerpost inside Helium is perfectly aligned with every centerpost of its surrounding other Heliums, making cubes.

These connected cubic patterns are the nervous systems of these assemblies, facilitating the transfer of temperature differences and/or energy transfers throughout all components.

This structure of 5 Heliums contains at 2 protons per Helium for a total of 10 protons and 10 neutrons, which is NEON.

In the next pages the proposed formation of all 8 of Period 2's elements will be shown.

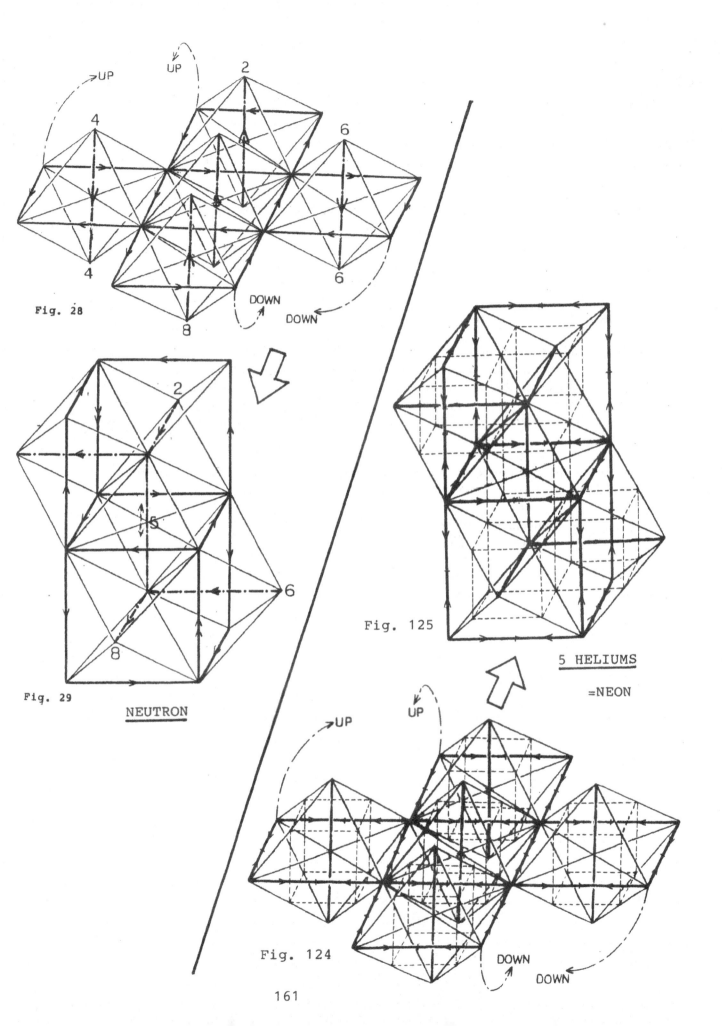

Fig. 28

Fig. 29

NEUTRON

Fig. 125

5 HELIUMS

=NEON

Fig. 124

161

Helium, with its 2 protons and 2 neutrons, is the fulfillment of Period 1 of the Periodic Table of Elements, in which the proton count decides its numerical place in the series of elements while the number of neutrons can be deduced from the atomic weight of each element.

In this theory Helium's perfect square shape and its 2 perfect 45° sloped bottom to bottom pyramids lend themselves ideally to be stacked in cubic patterns. Starting out, I had of course no idea what that assembly pattern would look like, but symmetry is a requirement.

While Helium is in this theory the large building block that will construct all seven periods of the Periodic Table, it is Helium's protons and neutrons that will be its smaller components that will assemble the Helium building blocks as this construction continues.

As an example: Period 2 starts with $^3Lithium^7$, which means it has 3 protons and 4 neutrons, so that one proton and 2 neutrons have been added after Helium. We might also say that Lithium has an extra Helium that has one less proton. Putting it that way, it is easier to comprehend what is actually happening. Next comes $^4Be^9$, which has gotten one more proton and one more neutron than Lithium, whereby this newly added proton has filled in the gap of the previous Helium that was lacking one proton.

The reality however also is that Lithium also can be created by adding at 2 locations of the original Helium a new Helium start-up, around the corner from each other.

And so it goes, making more elements, Fig. 126a, b, c, d, e, as Heliums construct the pre-ordained patterns of the Periodic Table.

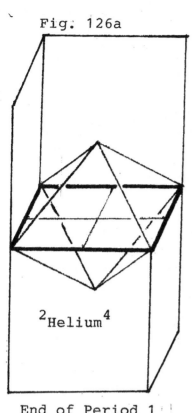

Fig. 126a

^{2}Helium^4

End of Period 1

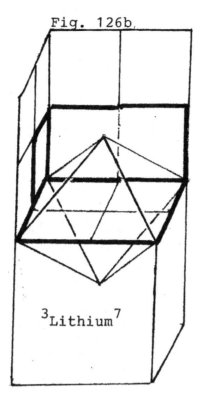

Fig. 126b

^{3}Lithium^7

Start of Period 2

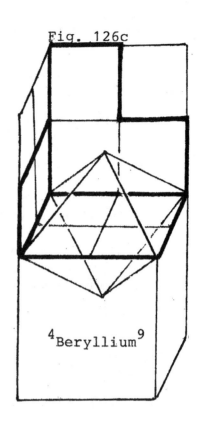

Fig. 126c

^{4}Beryllium^9

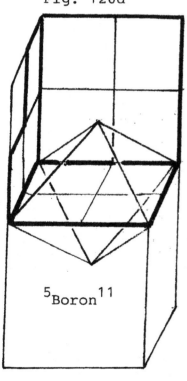

Fig. 126d

^{5}Boron^{11}

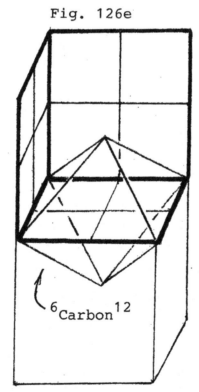

Fig. 126e

^{6}Carbon^{12}

Carbon is made of 3
complete heliums, which
are packaged in a tight
and compact assembly.
More on the next pages.

Carbon's perimeter is a

HEXAGON.

NOTE: The large bold-lined squares represent helium, which is shown here
      as being subdivided into 4 smaller squares, which are its 2 protons
      and 2 neutrons. At this time it has not yet been sorted out which
      are the protons or neutrons, which is why they are left unidentified.

Carbon is probably the most recognized element around us. Its versatility is noticeable very much in our lives when we see the trees, grasses, and the vegetables that we eat, having become aware of its immense versatility.

Most recently carbon-based very strong man made construction materials make great strides, as does the use of carbon in computer chips.

The reasons for these advances are no doubt not only due to its chemical characteristics, but by simultaneous being an element with a remarkably hexagon structure, shown on Fig. 126e, that lends itself to a self packaging method that has very useful structural and electronic consequences.

Enclosed front page of Science News of Sept. 29, 2007 shows any all-carbon electronic application that is being worked on in research centers. It was published at a perfect time as far as this theory is concerned.

I have built models of the carbons in this theory as they would look when they are made of 3 Heliums. The three carbons I built fit together beautifully.

SEPTEMBER 29, 2007   PAGES 193-208   VOL. 172, NO. 13

SCIENCE NEWS

THE WEEKLY NEWSMAGAZINE OF SCIENCE

the risk from plastics
mammoth hair's trove of dna
lifesaving mosquito nets
kelp's tropical locale

www.sciencenews.org

carbon circuits

Fig. 127

165

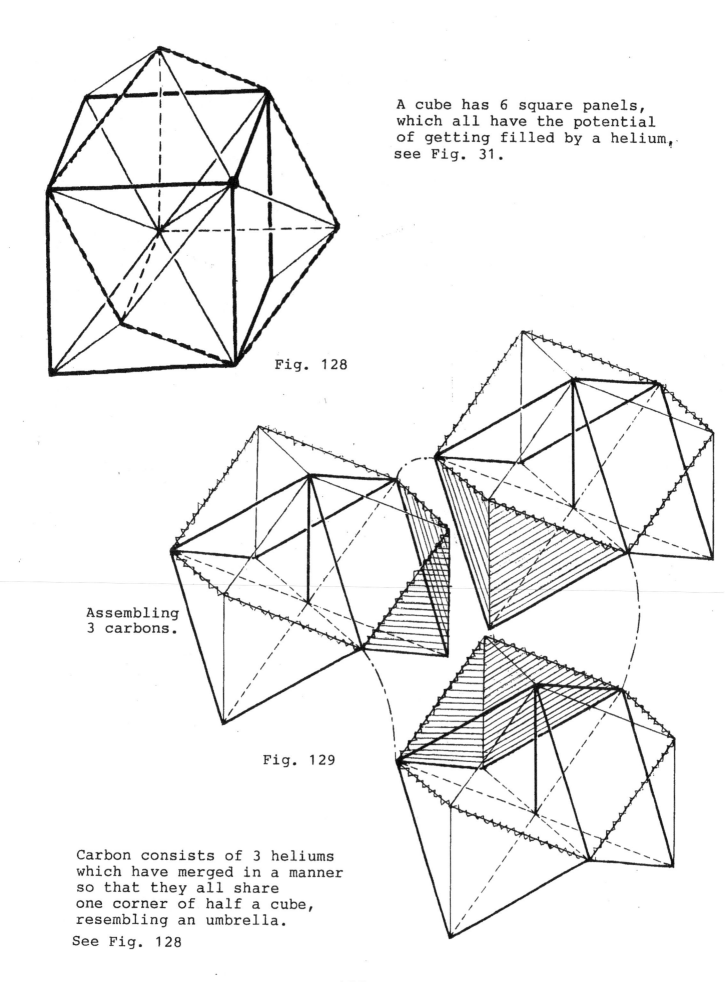

A cube has 6 square panels,
which all have the potential
of getting filled by a helium,
see Fig. 31.

Fig. 128

Assembling
3 carbons.

Fig. 129

Carbon consists of 3 heliums
which have merged in a manner
so that they all share
one corner of half a cube,
resembling an umbrella.

See Fig. 128

166

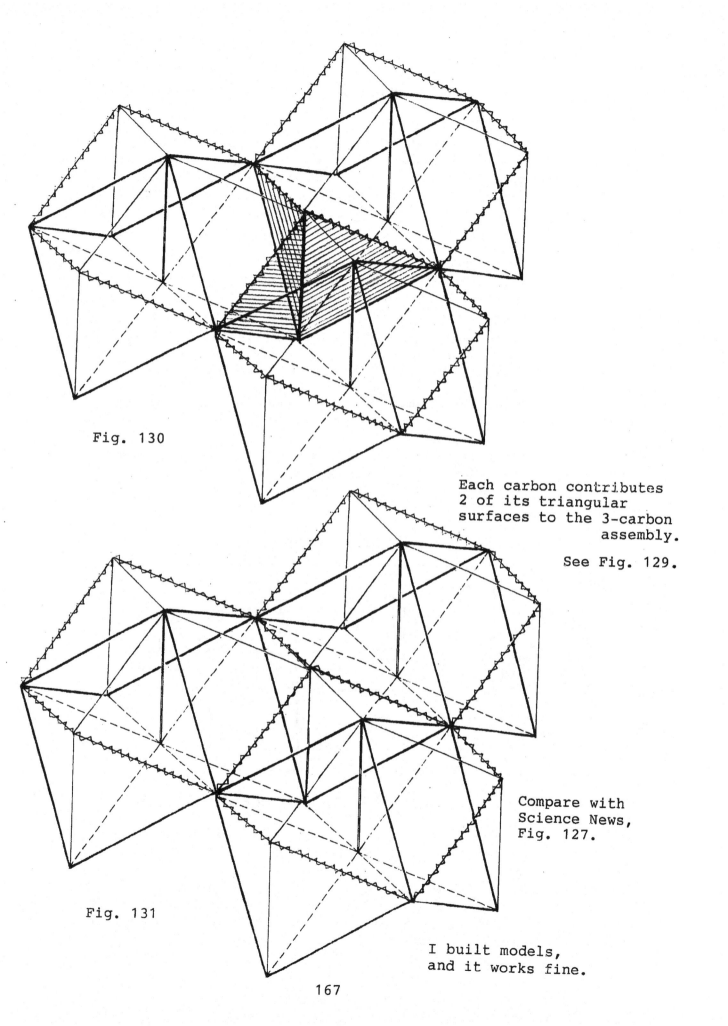

Fig. 130

Each carbon contributes
2 of its triangular
surfaces to the 3-carbon
assembly.

See Fig. 129.

Compare with
Science News,
Fig. 127.

Fig. 131

I built models,
and it works fine.

The next element, Nitrogen, is a gas which has a very low melting point. What happened here is that nature's need for symmetry with regards to the original Helium came into play, which was done by partly dismantling 2 of Carbon's 3 Heliums, and transferring those parts to the original Helium's other side, while simultaneously adding an electron and a proton to this revised assembly, thus transforming $^6C^{12}$ into a perfect diametrically balanced $^7Nitrogen^{14}$. This total remake of this compound Helium structure resulted in the transformation of the solid Carbon structure into a stable gas, Nitrogen, Fig. 132a.

Oxygen, following the same structural path as nitrogen, is diametrically perfect, and is a gas. Its Fig. 132b shows on both ends of sheltered opening which one might see as the jaws of a shark that can hang onto its prey, thus explaining its aggressiveness in oxidizing other elements. Oxygen's 2 tight sheltered unfilled openings are perfect sites to harbor a single Hydrogen, making Hydroxyl OH.

Fluor has a little less symmetry, as compared with Nitrogen and Oxygen, but it also is a gas, Fig. 132c. The last element in the row of Period 2 is Neon, Fig. 132d, which consists of the original Helium panel that now has been joined in a diametrically perfect construction method by 4 additional Heliums.

And viola! Here nature has made an element that is so structurally perfect as a gas that it is inert to its surroundings, thereby nearly succeeding in its efforts to return to the womb, which was before the Big Bang, doing that by recreating an identical shape as the neutron, nature's original particle in this theory, but on a larger scale, Fig. 133. →

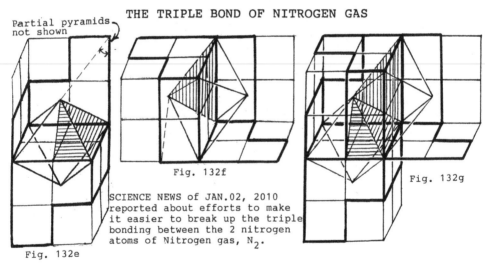

Partial pyramids not shown

THE TRIPLE BOND OF NITROGEN GAS

Fig. 132f

Fig. 132g

SCIENCE NEWS of JAN.02, 2010 reported about efforts to make it easier to break up the triple bonding between the 2 nitrogen atoms of Nitrogen gas, N_2.

Fig. 132e

That story prompted the effort here in this theory to find out if these bonds could be found in this Decoding theory.

To find out I built two cardboard nitrogen atoms as seen in Feb. 132a, and then tried to put them together in several ways, looking for the triple bond, and it took less than 15 minutes to do it. The solution:

N-atom #1 of Fig.132e gets its cubic upper half invaded by the cubic upper half of N-atom #2 of Fig. 132f which has been laid down and rotated 90°. The result is Fig. 132g which shows that both nitrogen atoms have merged by using their identical 2 full whole pyramid slopes and one ½ pyramid slope. This system works. (For 'CHEMICAL BONDING' see page 106.)

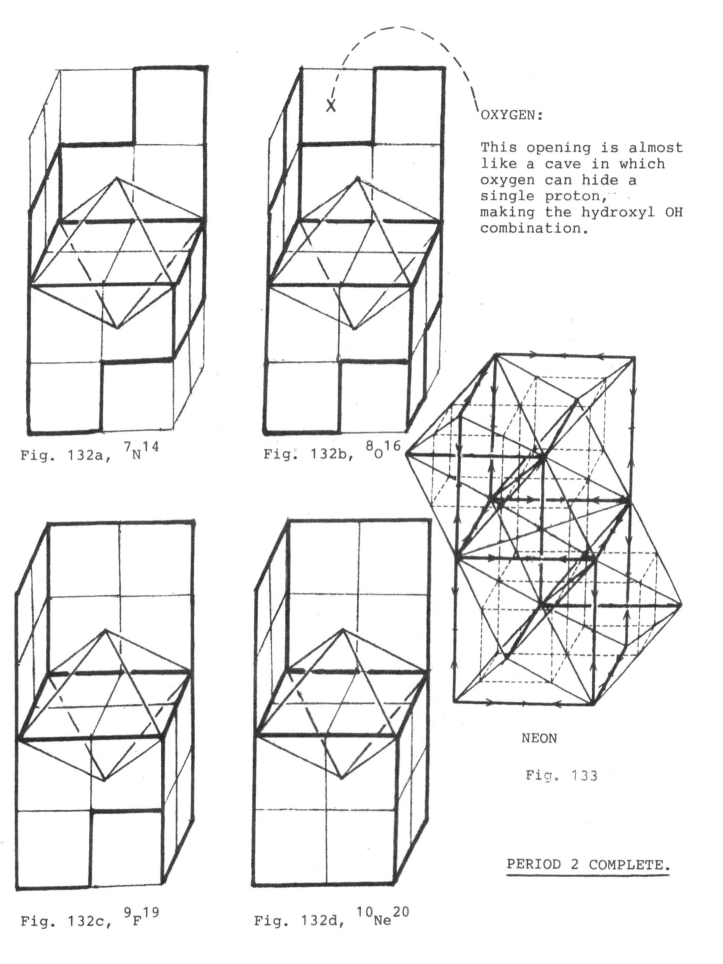

Fig. 132a, $_7N^{14}$

Fig. 132b, $_8O^{16}$

OXYGEN:

This opening is almost
like a cave in which
oxygen can hide a
single proton,
making the hydroxyl OH
combination.

NEON

Fig. 133

PERIOD 2 COMPLETE.

Fig. 132c, $_9F^{19}$

Fig. 132d, $_{10}Ne^{20}$

Having completed Period 2 with the formation of NEON as shown on Fig. 133, for convenience it is shown here again as Fig. 134 to facilitate the explanation of how Argon's Period 3 adds on to Neon's Period 2.

Period 2 and Period 3 are like identical twins, they both have a sequence that attaches itself to the initial Hyrogen of Period 1 by constructing the 8 new elements in their series, one by one, with the help of 8 neutrons.

In doing so, this was done by gradually adding 2 heliums to the helium that was Period 1, thereby making carbon, which then occupies a half-cube. In the second half of Period 2 an identical half-cube was added to the other side of the initial helium, making neon.

Period 3 completes this cube-building process by filling in the initial helium's 5 still unoccupied spaces. These 4 open spaces are then gradually in a symmetrical way filled by the in Fig. 135 shown square ring of 4 heliums, producing the completed Argon of Fig. 136.

Fig. 137 is Argon's basic skeleton which will be duplicated in the formation of Krypton, Xenon and Radon.

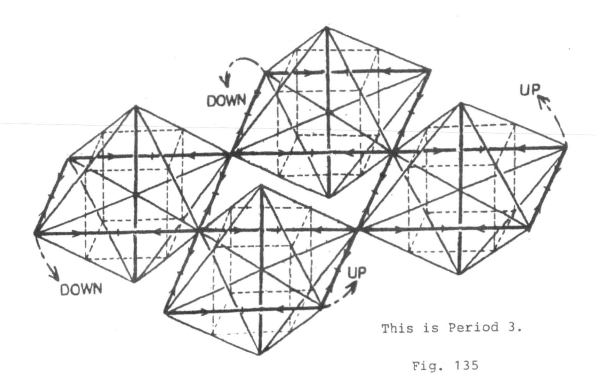

This is Period 3.

Fig. 135

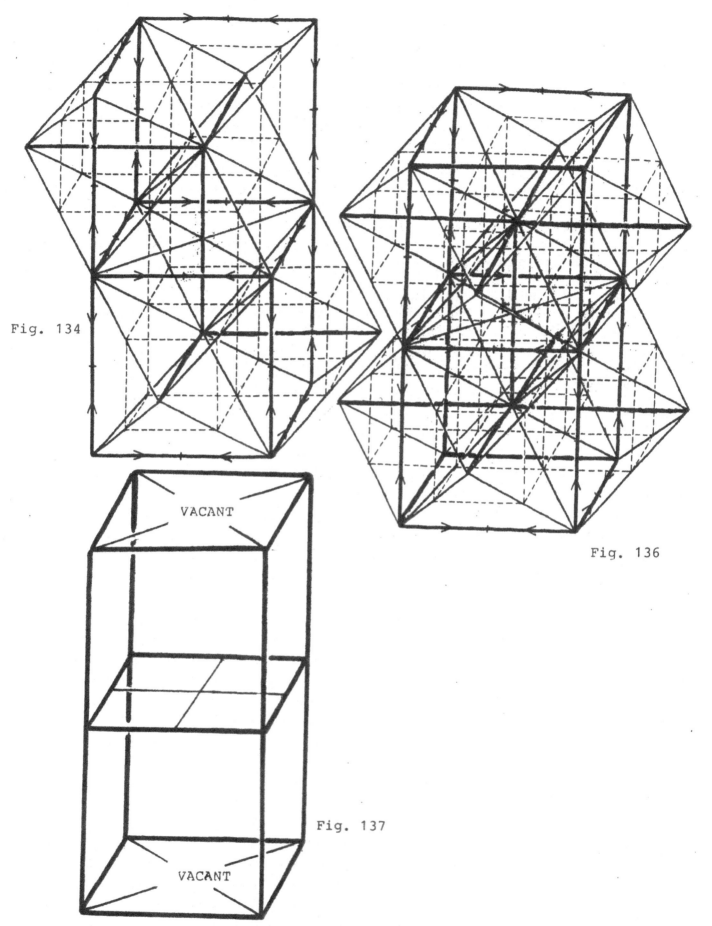

Fig. 134

Fig. 136

VACANT

VACANT

Fig. 137

171

With the creation of Argon, we have now arrived at the notion that Argon's shape represents the standard major assembly that will serve as the first cohesive prefabricated unit that will from hereon be repeated four more times during the final assembly of the Periodic Table.

The dominant opinion of scientists at this time is that the Periodic Table is an expression of an assembly of ever increasing concentric spheres in which electrons are orbiting a cluster of protons and neutrons in the sphere's center. In this theory that is not the case, in which two proton/neutron pairs are packaged, making the element Helium, which then becomes the building block for Argon's POD-shaped structure of 9 Heliums, which is made of a tight assembly of Periods 1, 2 and 3, which had a growth from 1 to 18 elements by its protons and neutrons.

Periods 4 and 5 both also have 18 elements, and one can recognize the (2+6) "A" group elements that imitate periods 2 and 3, and also the (8+2) "B" group that has those 2 extra elements, which is their way of accounting for Period 1's Helium.

Period 4, Krypton, and Period 5, Xenon are thus matching Argon's 18-element Pods, see Fig. 138 and 139.

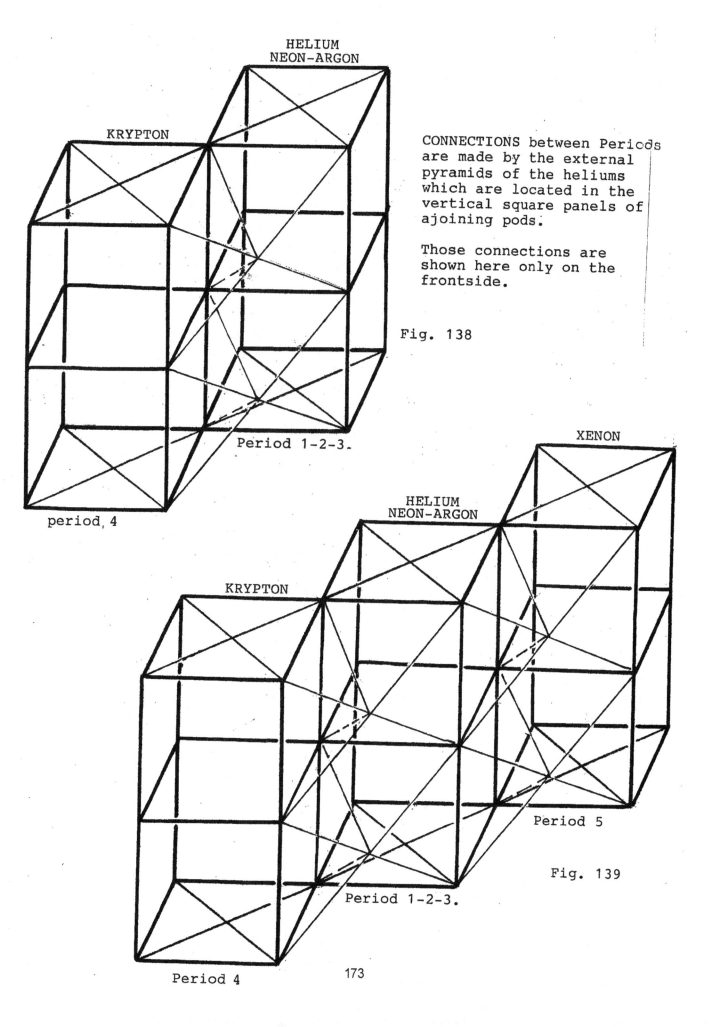

KRYPTON

HELIUM
NEON-ARGON

CONNECTIONS between Periods
are made by the external
pyramids of the heliums
which are located in the
vertical square panels of
ajoining pods.

Those connections are
shown here only on the
frontside.

Fig. 138

Period 1-2-3.

period 4

XENON

HELIUM
NEON-ARGON

KRYPTON

Period 5

Period 1-2-3.

Fig. 139

Period 4

173

The growth of the Periodic Table, Fig. 134, 135, 136, 137 for Argon, Fig. 138 for Krypton and Fig. 139 for Xenon has indicated the growth sequence of Periods 3, 4 and 5 as they become 3 PODS in a row, which connect with their protruding pyramids at one of their corners with each other.

When Argon continues with the Periodic Table's growth, it has a choice to do that at one of its corners of choice, but it might just as well have chosen the corner that is opposite its choice. For the sake of symmetry, it is very likely that that Argon at first will grow at 2 opposite sides for the sake of symmetry, but it is suspected that with ^{26}Iron it has made a perfectly balanced left and right that lends itself to producing magnetism, as explained on Fig. 140.

When continuing to grow after Iron with Cobalt and Nickel in this Period 4, it had already gained 10 new elements in this Period 4, which was enough to transform it to a Neon-like 9 unit structure on one side. It seems that instead of having a symmetrical left-right structural balance attached to its underlying Period 3, that it now newly made packaged 9-unit inert Neon structure was a better liked arrangement that reminds one of the Carbon-Nitrogen remake.

At this point Period 3's Pod has a single growing half-pod at one side which will then grow to the full pod that will be Krypton of Period 4, thus imitating the growth that had produced Argon in Period 3.

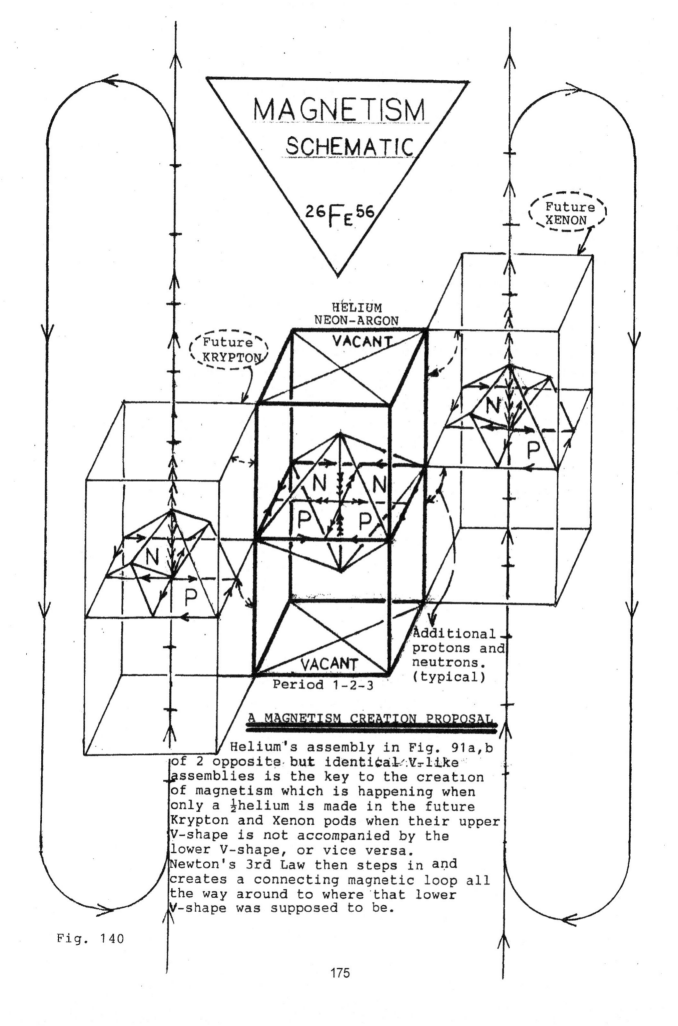

MAGNETISM
SCHEMATIC

$^{26}Fe^{56}$

HELIUM
NEON-ARGON
VACANT

Future
KRYPTON

Future
XENON

N N
P P

N
P

N
P

VACANT
Period 1-2-3

Additional
protons and
neutrons.
(typical)

A MAGNETISM CREATION PROPOSAL

   Helium's assembly in Fig. 91a,b
of 2 opposite but identical V-like
assemblies is the key to the creation
of magnetism which is happening when
only a ½helium is made in the future
Krypton and Xenon pods when their upper
V-shape is not accompanied by the
lower V-shape, or vice versa.
Newton's 3rd Law then steps in and
creates a connecting magnetic loop all
the way around to where that lower
V-shape was supposed to be.

Fig. 140

175

The growth of the Periodic Table for the past century from Hydrogen and Helium onward seems always to have been on the basis of every increasing and larger concentric spheres around Helium.

Period 2's Neon and Period 3's Argon are very much alike, as if they were twins, and seem to have a rather simple spatial structure.

Period's 4 Krypton and period's 5 Xenom are also very much alike to each other, but their own sequence of 18 elements consists of a pair of intertwined 8-unit "A" and "B" groups that resemble Periods 2 and 3, plus 2 extra "VIIIB" elements that take the place of Helium's Period 1. It gets complicated.

Following those complicated structures of Krypton and Xenon, here comes Radon's Period 6 with an 18-unit duplicate of Krypton and Xenon periods, but in addition to that it has a 14-unit element group, the Lanthanum Group, which seems to be odd and difficult to find a place for in the adopted system of concentric spheres.

The POD-system as described herein provides a prefect fit for the above described periods, and Period 7 will fit in it as well.

Taking for a moment a few steps ahead of the yet to be described Radon and beyond, Fig. 141 is the structure that houses all of the completed Periodic Table.

In this theory, the totally complete Periodic Table consists of a stack of two 3x3 layers of vacant cubes which has at its center a single Helium. Overall there are 18 cubic spaces.

There are 75 square panels in the entire assembly, all of them suitable to hold a Helium.

Helium's atomic weight is 4.0000; if all of its 75 panels had Helium in it, its total atomic weight would be 75x4.0000 = 300.0000.

THIS IS THE COMPLETE PERIODIC TABLE'S STRUCTURAL FRAME

Fig. 141

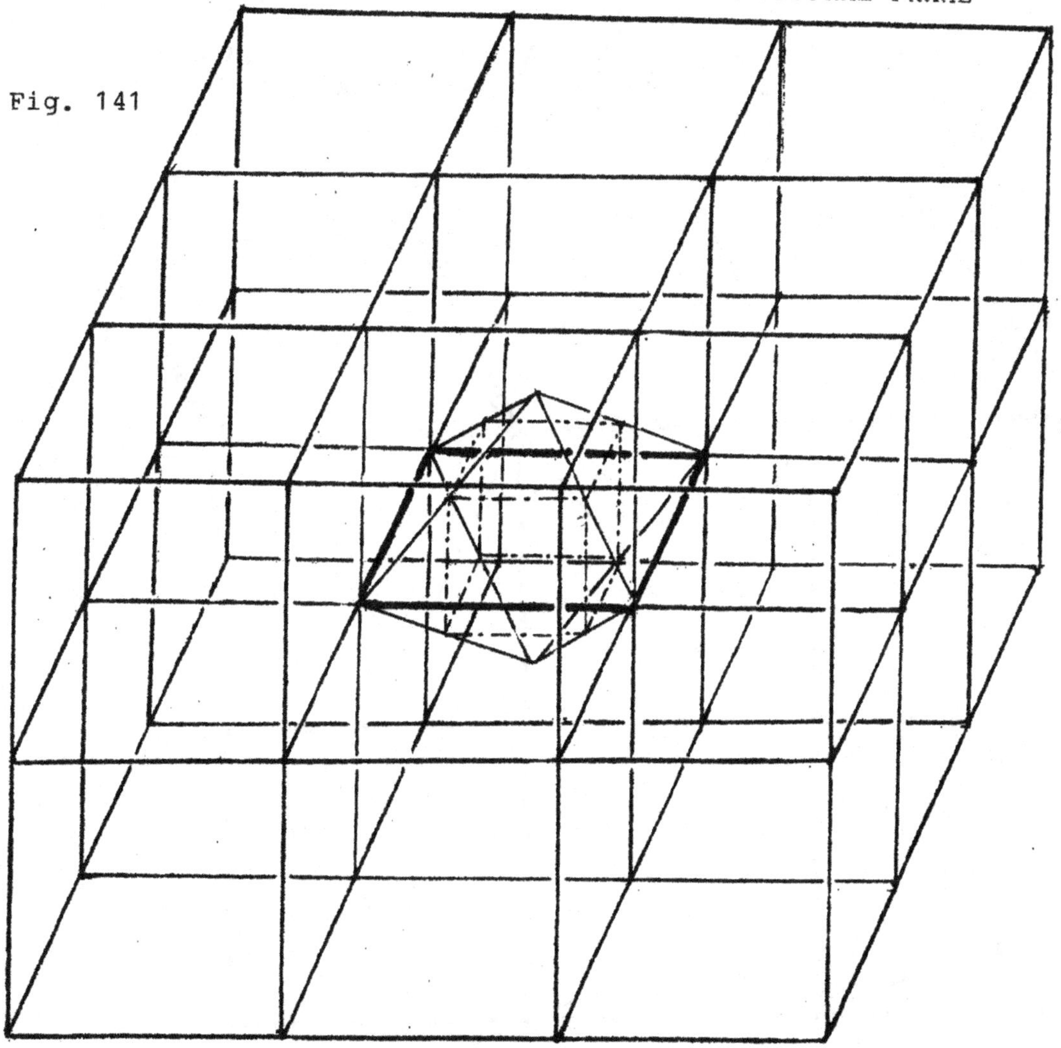

Particles such as the neutron, proton and elecron are encapsulated in a cloud-like spatial frame as shown above that facilitates the joining of those particles into larger structures by providing alignments of the spatial matrix's squares.

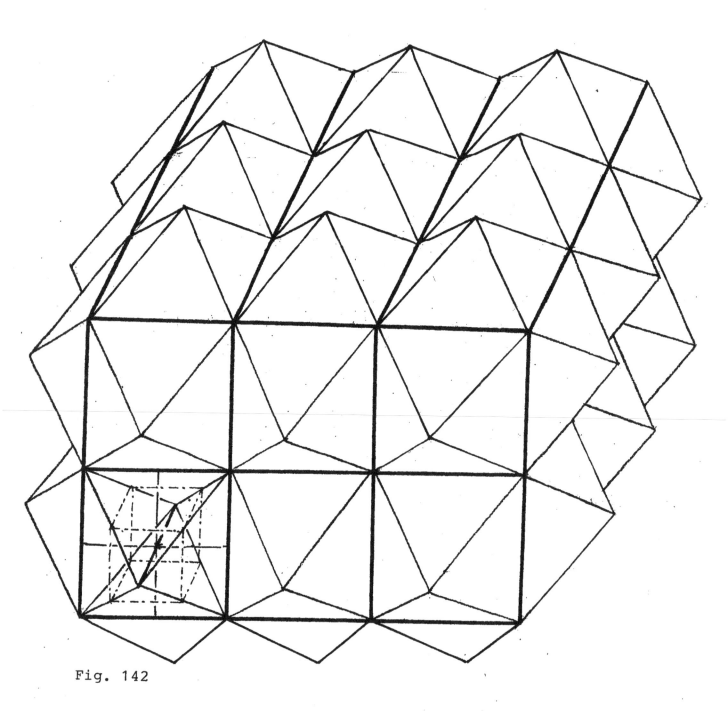

Fig. 142

THE COMPLETE PERIODIC TABLE
WITH ALL OF ITS HELIUMS IN PLACE

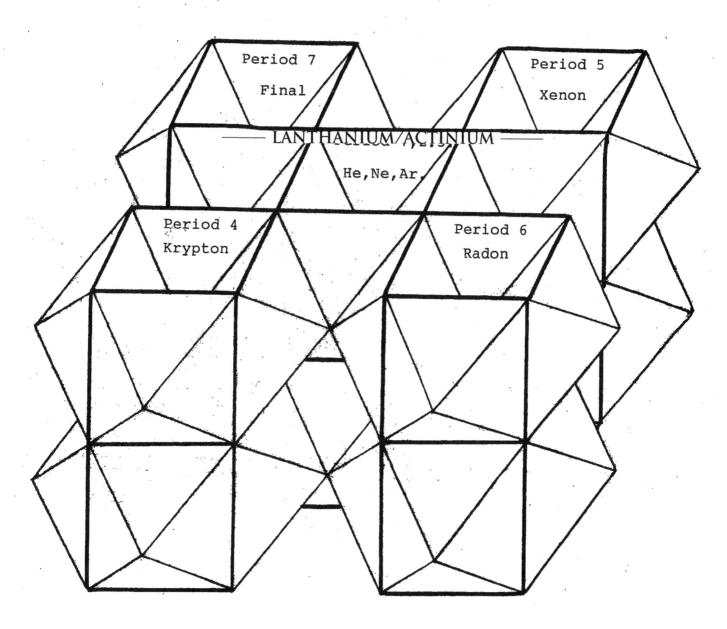

Fig. 143

A CUT-AWAY DRAWING OF HOW THE
PERIODIC TABLE'S FIVE PODS CONNECT

Period 6's Argon-like pod at the lower right has 18 protons
in its 9 Heliums. The top and bottom of that pod have no Helium
in it, just like Periods 3, 4 and 5.

It is suspected in this theory that Period 6's Lanthanium Group's
14 elements can be found
in the 7 Heliums of the
roof panels of the
Periodic Table's assembly.
(marked with an oval).

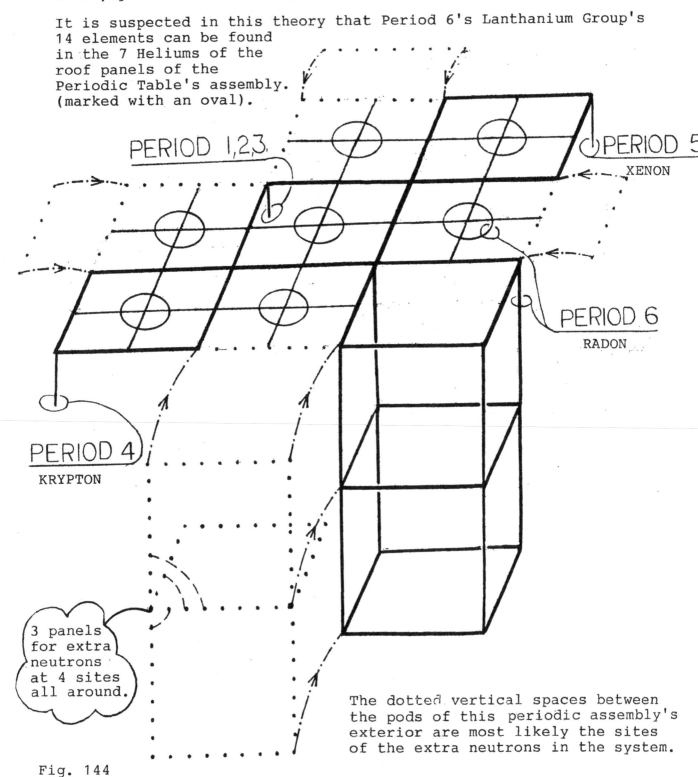

PERIOD 1,2,3.

PERIOD 5

XENON

PERIOD 6

RADON

PERIOD 4

KRYPTON

3 panels
for extra
neutrons
at 4 sites
all around.

The dotted vertical spaces between
the pods of this periodic assembly's
exterior are most likely the sites
of the extra neutrons in the system.

Fig. 144

Radon's Period 6 has 32 elements, and it is generally accepted that
Period 7 would have the same number, Fig. 145.
The last natural element is Uranium which is actually the fourth
element of the Actinium Group.

While working on trying to decipher the Periodic Table, I had noticed
the remodeling of Carbon into Nitrogen and also right after Iron's
magnetism. There is no doubt in this theory that that remodeling
is common practice throughout the Periodic Table.

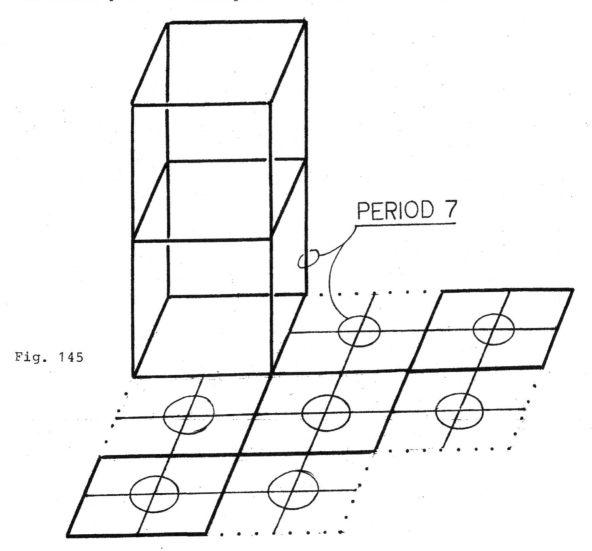

PERIOD 7

Fig. 145

Period 7 is the upside down version of Period 6, but it is incomplete.
It seems that there is something going on that makes it more difficult
to fill the still vacant spaces, whatever that may be.

Retroactively, nature sees an opportunity to build another argon-like
18-unit pod by stripping element Uranium's last 4 proton's of its
Actinium group away, and then by also stealing all 14 protons from
Period 6's Lanthanium group, it has enough prefabricated pieces on
hand to create Krypton and Barium, with no leftovers.

	Counting Protons and Neutrons		Total Panels Filled
PERIOD 1 – $^{2}\text{He}^{4}$	2	2	1
PERIOD 2 – $^{10}\text{He}^{20}$	10	10	5
PERIOD 3 – $^{18}\text{Ar}^{40}$	18	18+4	10
PERIOD 4 – $^{36}\text{Kr}^{84}$	36	36+12	21
PERIOD 5 – $^{54}\text{Xe}^{131}$	54	54+23	33
PERIOD 6 – $^{86}\text{Rn}^{222}$	86	86+50	55
PERIOD 7 – $^{118}\text{Fi}^{300}$	118	118+64	75 Full

Putting this theory together was greatly aided by using the number of elements in each period.

By adding the 2 Elements of Period 1 to the 8 elements of Period 2 and then by also adding the 8 elements of Period 3 makes for a total of 18 elements, equaling the 18 elements of Period 4 and also of Period 5. Periods 6 and 7, without their 14-unit Lanthanum and Actinium Groups respectively, are similar to Periods 4 and 5, thus having a total of five 18-unit groups plus the two 14-unit groups.

What also was very helpful in this search was the great similarity of periods Neon through Radon that their first two and last six elements were so very much alike. I was looking for road signs everywhere, and the discovery physicist by Lise Meitner in the early 1940's speaks volumes of what was going on with the various periods when Uranium falls apart into Krypton and Barium.

Fig. 146 shows a conventional Periodic Table of Elements which was originally conceived by Russian Chemist Dimitri I. Mendeleev in 1869. It is based on the chemical characteristics of its elements, but its overall appearance suggests also that it structurally represents concentric spheres from period to period.

In this theory a new Periodic Table is presented in Fig. 147 which is based on the STRUCTURAL configuration of the assembled elements.

AN OBSERVATION ABOUT RADIOACTIVITY:

In a comprehensive review of the various structures of all of the elements of the Periodic Table, it seems that the radioactivity of an element is caused by a serious lack of symmetry in its overall structure.

The affected particle then attempts to correct that issue by gradually ejecting those parts of its structure which by their absence would improve the remaining structural symmetry of the particle.

DECODING
THE PERIODIC TABLE

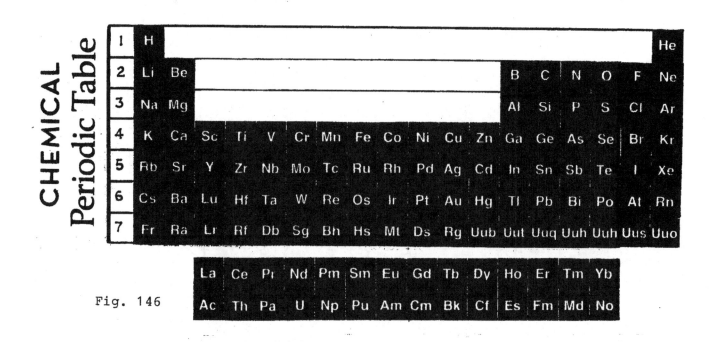

CHEMICAL Periodic Table

Fig. 146

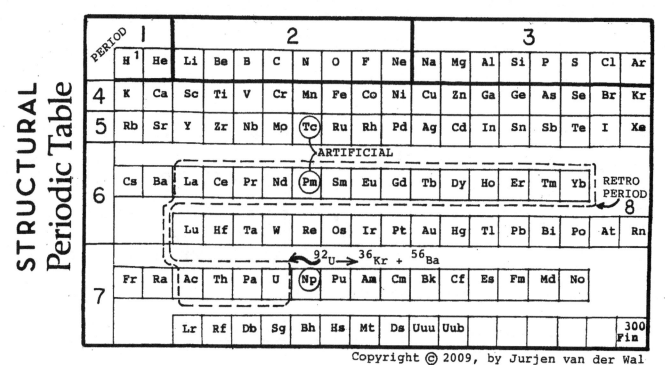

STRUCTURAL Periodic Table

Fig. 147

Jurjen van der Wal

183

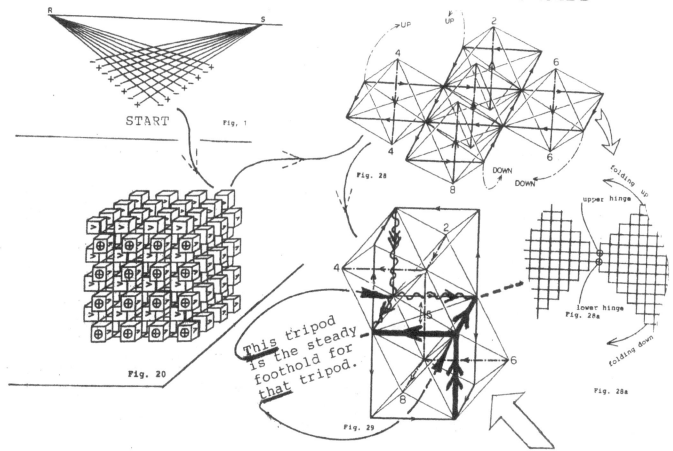

START   Fig. 1

Fig. 28

Fig. 28a

upper hinge

lower hinge
Fig. 28a

folding up

folding down

Fig. 20

This tripod is the steady foothold for that tripod.

Fig. 29

THE PACKAGING

OF

NUCLEAR MASS

In this theory it is a standard feature that the (boldly lined) basic tripod of Fig. 29 shown above splits its baseline pair with its 2 arrows into 2 single lines with one arrow each.

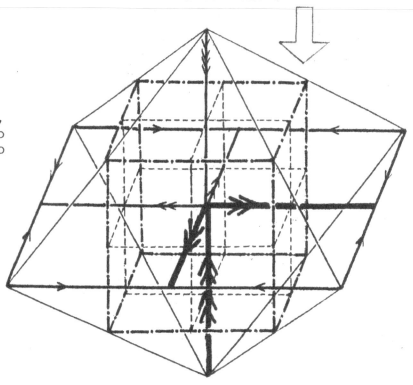

When Helium gets constructed as shown in Fig. 107,108,109, it is doing so by pairing two basic tripods of Fig. 29 into a single (boldlined) tripod with twice the arrows.

Shown here for clarity, is another tripod, upside down, but not outlined boldly, located above the other one.

Helium,   Fig. 148

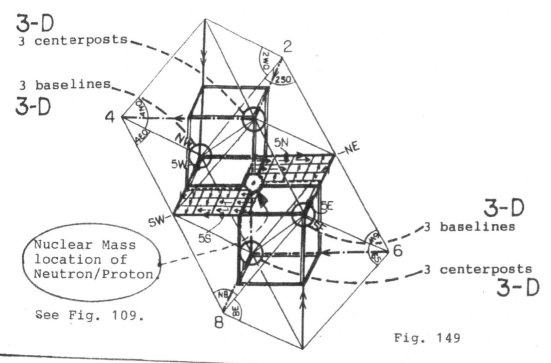

3-D
3 centerposts

3 baselines
3-D

Nuclear Mass
location of
Neutron/Proton

See Fig. 109.

3-D
3 baselines

3 centerposts
3-D

Fig. 149

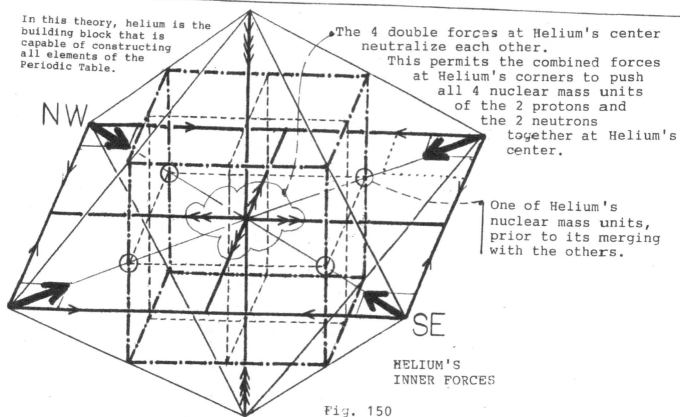

In this theory, helium is the
building block that is
capable of constructing
all elements of the
Periodic Table.

The 4 double forces at Helium's center
neutralize each other.
This permits the combined forces
at Helium's corners to push
all 4 nuclear mass units
of the 2 protons and
the 2 neutrons
together at Helium's
center.

One of Helium's
nuclear mass units,
prior to its merging
with the others.

HELIUM'S
INNER FORCES

Fig. 150

All forces within the flat, square, shared baseplate between the upper
and lower pyramids are in balance with each other. The bold arrows from
the four corners of the baseplate are exerting a compressive force
towards Helium's inner center.

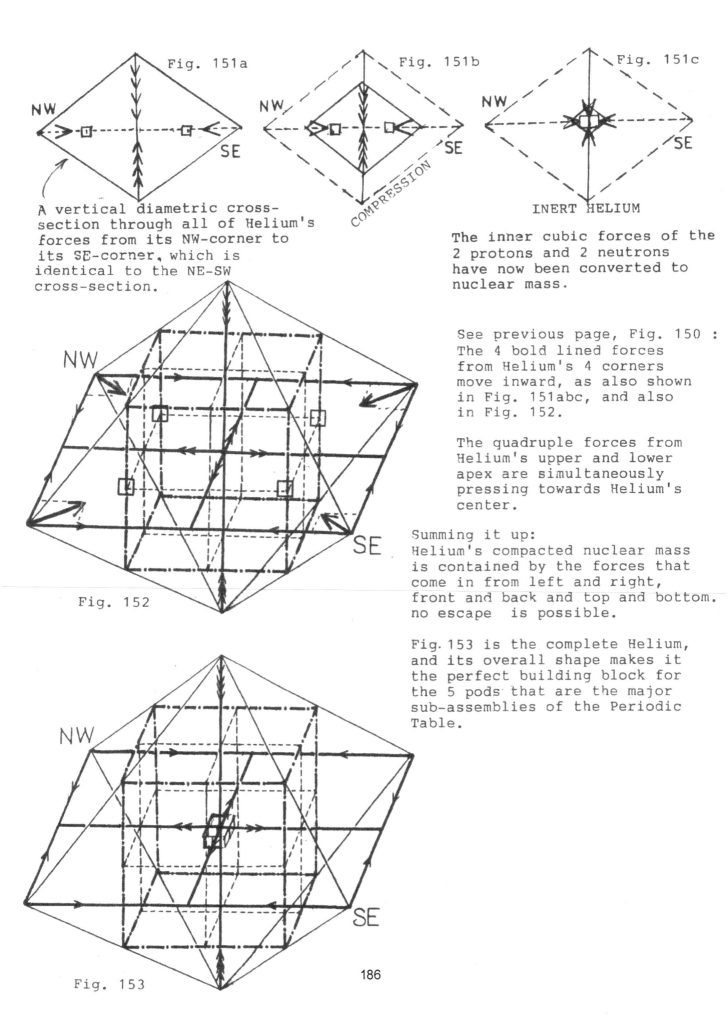

Fig. 151a

Fig. 151b

Fig. 151c

NW

SE

NW

SE

NW

SE

COMPRESSION

INERT HELIUM

A vertical diametric cross-section through all of Helium's forces from its NW-corner to its SE-corner, which is identical to the NE-SW cross-section.

The inner cubic forces of the 2 protons and 2 neutrons have now been converted to nuclear mass.

NW

SE

Fig. 152

See previous page, Fig. 150 : The 4 bold lined forces from Helium's 4 corners move inward, as also shown in Fig. 151abc, and also in Fig. 152.

The quadruple forces from Helium's upper and lower apex are simultaneously pressing towards Helium's center.

Summing it up:
Helium's compacted nuclear mass is contained by the forces that come in from left and right, front and back and top and bottom. no escape is possible.

Fig. 153 is the complete Helium, and its overall shape makes it the perfect building block for the 5 pods that are the major sub-assemblies of the Periodic Table.

NW

SE

Fig. 153

186

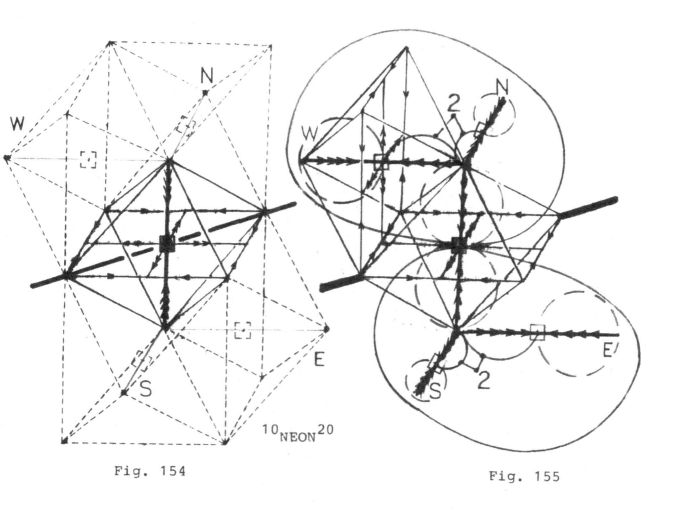

Fig. 154

Fig. 155

$10_{NEON}20$

The 5 Heliums that make Neon have an assembly in which the central
Helium is in a symmetrical way surrounded by 4 other Heliums, Fig. 154.
As this theory develops, we see in Fig. 155 that every centerpost of
Neon's 5 Heliums has 2 sets of 4 arrows which have captured the combined
nuclear mass of its 2 protons a and 2 neutrons between them.

A thick heavy line splits Neon into 2 identical halves, which then look
like 2 identical tripods. Each tripod has 3 (circled) quadruple arrow
groups which point toward Helium's center, and it also has 2 (half-circled)
quadruple arrow groups which are pointing away from Neon's center.

The 3 quadruple arrow groups are of course stronger than the 2 quadruple
arrow groups, and as a consequence the 4 captured Helium nuclear mass
units from the N,W,S and E tripod legs are pushed towards Neon's center,
where all 4 of them join Neon's single central Helium, making a solid
cluster of Neon's nuclear mass.

This arrangement allows unhindered passage through the vacant centerposts
of all protons and neurons of this entire Neon structure, which comes
into play when influances from the outside such as temperature changes
are equally distrbuted throughout the entire Neon structure.

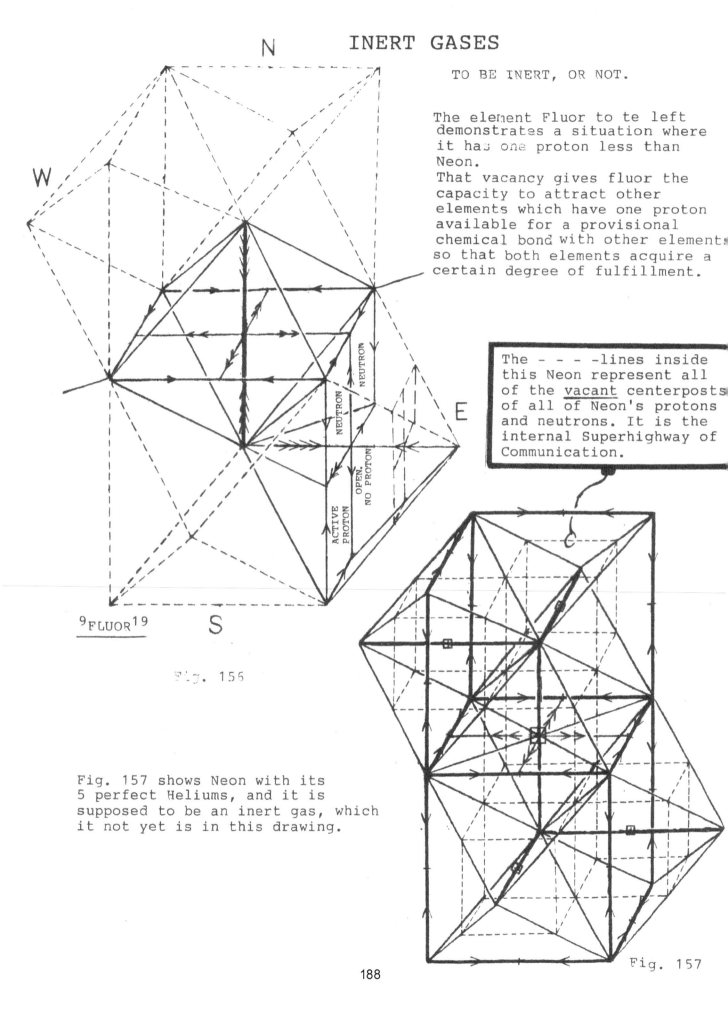

TO BE INERT, OR NOT.

The element Fluor to te left demonstrates a situation where it has one proton less than Neon.
That vacancy gives fluor the capacity to attract other elements which have one proton available for a provisional chemical bond with other elements so that both elements acquire a certain degree of fulfillment.

The - - - -lines inside this Neon represent all of the vacant centerposts of all of Neon's protons and neutrons. It is the internal Superhighway of Communication.

N

W

E

S

^{9}FLUOR^{19}

Fig. 156

NEUTRON
NEUTRON
NEUTRON
ACTIVE PROTON
OPEN. NO PROTON
NO PROTON

Fig. 157 shows Neon with its 5 perfect Heliums, and it is supposed to be an inert gas, which it not yet is in this drawing.

Fig. 157

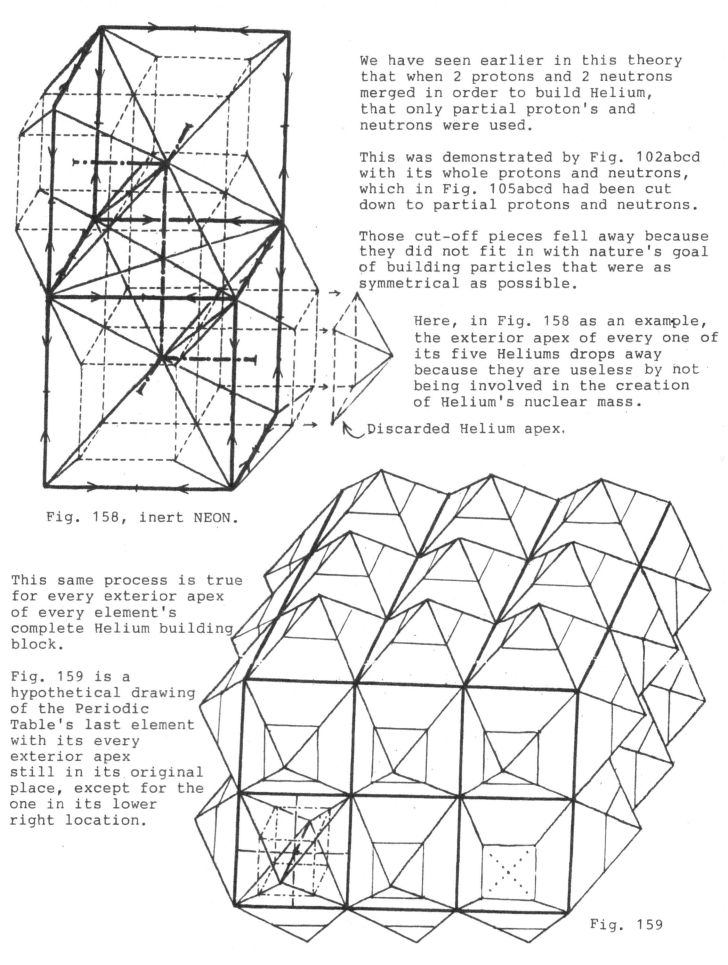

We have seen earlier in this theory
that when 2 protons and 2 neutrons
merged in order to build Helium,
that only partial proton's and
neutrons were used.

This was demonstrated by Fig. 102abcd
with its whole protons and neutrons,
which in Fig. 105abcd had been cut
down to partial protons and neutrons.

Those cut-off pieces fell away because
they did not fit in with nature's goal
of building particles that were as
symmetrical as possible.

Here, in Fig. 158 as an example,
the exterior apex of every one of
its five Heliums drops away
because they are useless by not
being involved in the creation
of Helium's nuclear mass.

Discarded Helium apex.

Fig. 158, inert NEON.

This same process is true
for every exterior apex
of every element's
complete Helium building
block.

Fig. 159 is a
hypothetical drawing
of the Periodic
Table's last element
with its every
exterior apex
still in its original
place, except for the
one in its lower
right location.

Fig. 159

189

LISE MEITNER'S LEGACY

This Fig. 160 shows the 2 locations where in this theory the elements are located which provide the protons and neutrons that are needed to construct retro Period 8, which is that 18-unit pod that enables ^{92}Uranium to break up into ^{36}Krypton and ^{56}Barium.

Uranium's break-up was discovered by Lise Meitner, born in 1878, she was an Austrian nuclear physicist active in Sweden.

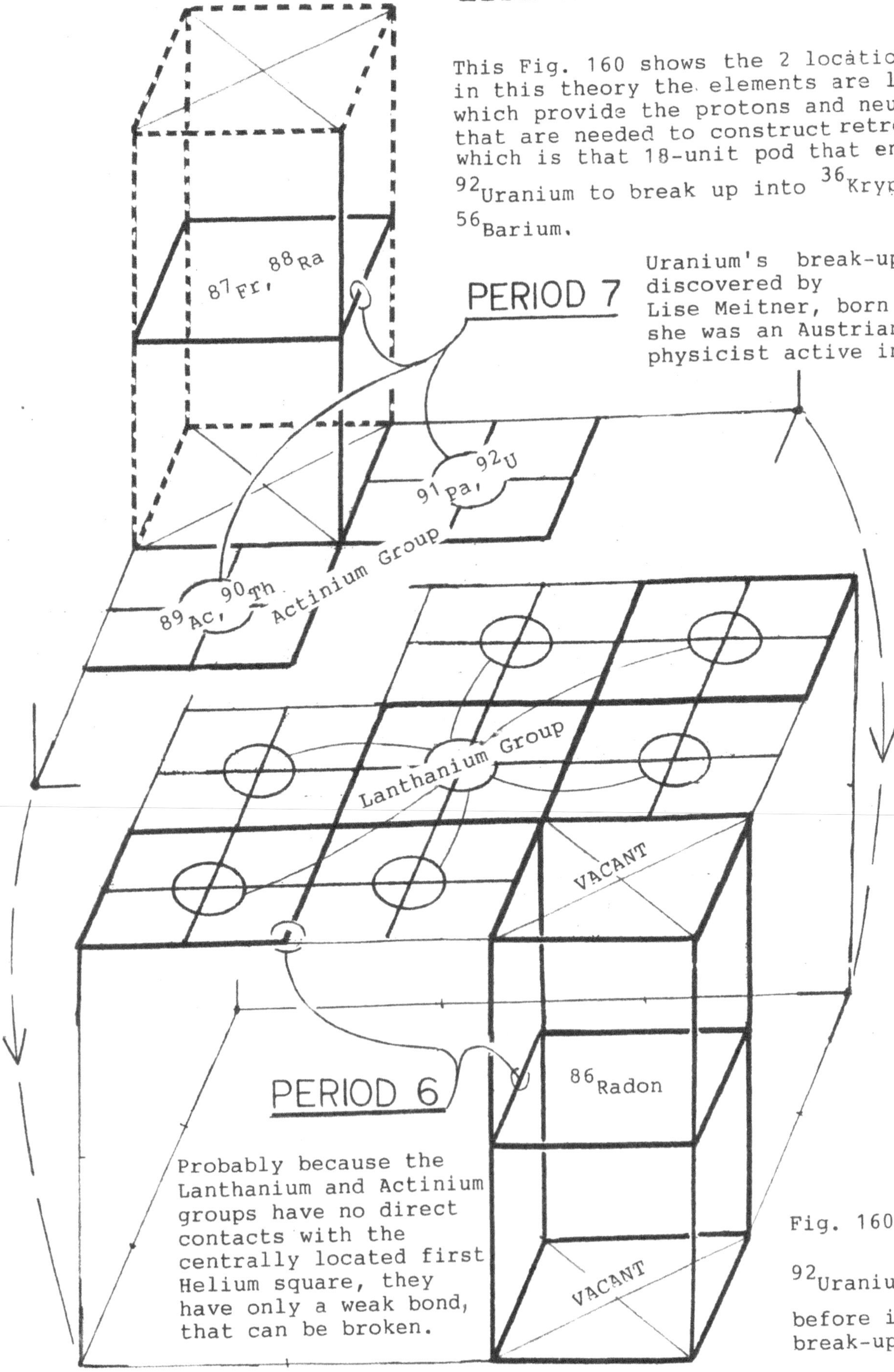

^{87}Fr, ^{88}Ra

PERIOD 7

^{91}Pa, ^{92}U

Actinium Group

^{89}Ac, ^{90}Th

Lanthanium Group

VACANT

PERIOD 6

^{86}Radon

Probably because the Lanthanium and Actinium groups have no direct contacts with the centrally located first Helium square, they have only a weak bond, that can be broken.

VACANT

Fig. 160

^{92}Uranium

before its break-up.

190

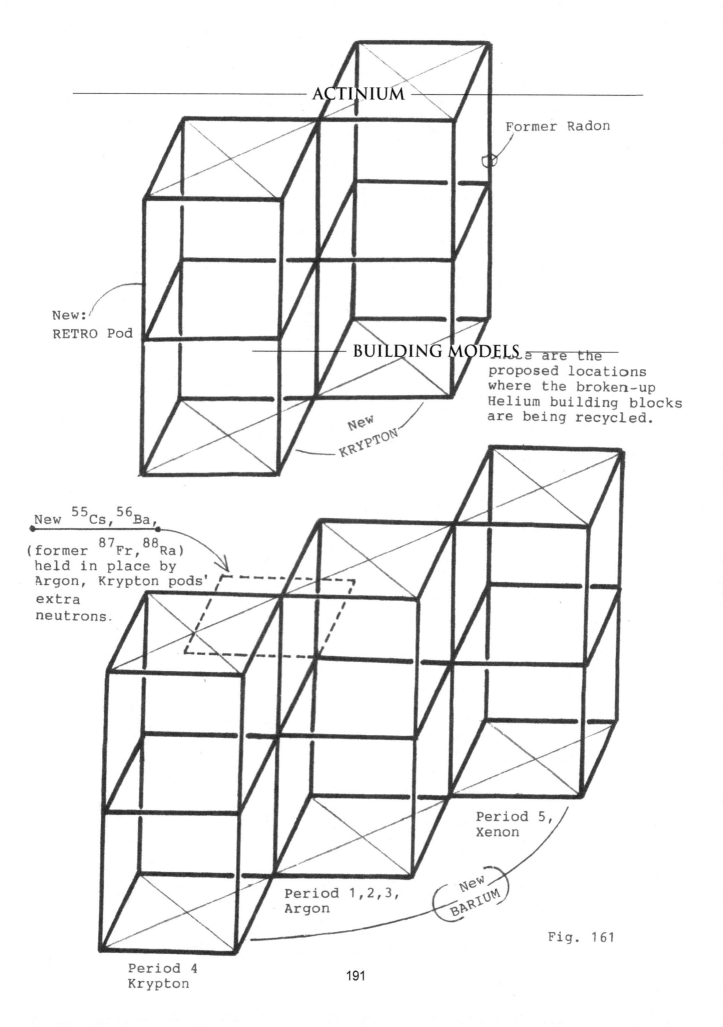

ACTINIUM

Former Radon

New:
RETRO Pod

BUILDING MODELS ...e are the
proposed locations
where the broken-up
Helium building blocks
are being recycled.

New
KRYPTON

New ^{55}Cs, ^{56}Ba,
(former ^{87}Fr, ^{88}Ra)
held in place by
Argon, Krypton pods'
extra
neutrons.

Period 5,
Xenon

Period 1,2,3,
Argon

New
BARIUM

Fig. 161

Period 4
Krypton

191

BUILDING MODELS

It will be of great benefit to the reader to have actual models in his or her hands while reading this Theory.
Shown here are the patterns which I used for the models.
Use a thin but stiff cardboard or plastic.

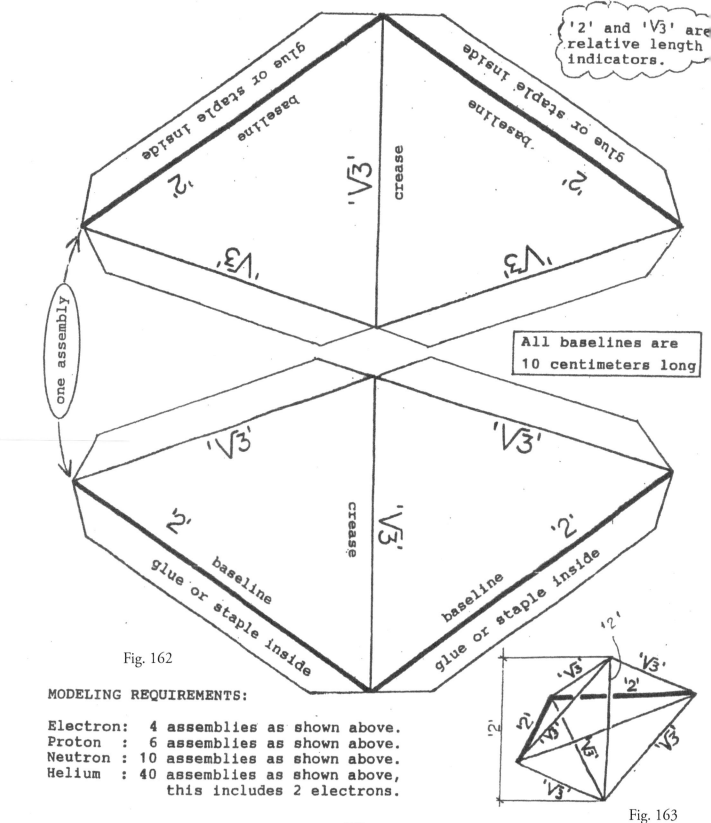

'2' and 'V3' are relative length indicators.

All baselines are 10 centimeters long

one assembly

Fig. 162

Fig. 163

MODELING REQUIREMENTS:

Electron: 4 assemblies as shown above.
Proton : 6 assemblies as shown above.
Neutron : 10 assemblies as shown above.
Helium : 40 assemblies as shown above,
              this includes 2 electrons.

SCAR COVERS

(Where ¼ electrons were attached to proton.)

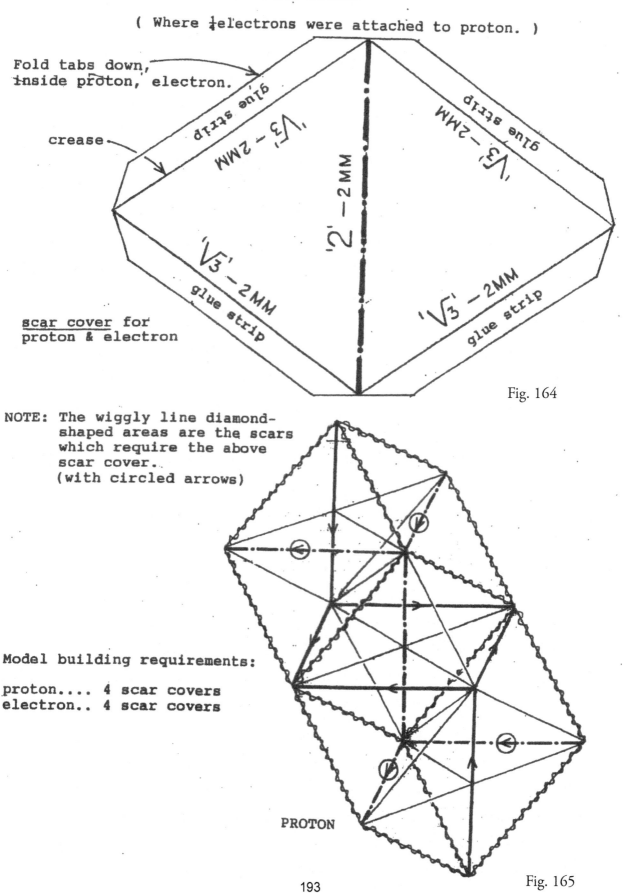

Fold tabs down,
inside proton, electron.

crease

glue strip

'$\sqrt{3}$' − 2MM

'2' − 2MM

'$\sqrt{3}$' − 2MM

glue strip

scar cover for
proton & electron

'$\sqrt{3}$'− 2MM

'$\sqrt{3}$' − 2MM

glue strip

glue strip

Fig. 164

NOTE: The wiggly line diamond-
shaped areas are the scars
which require the above
scar cover.
(with circled arrows)

Model building requirements:

proton.... 4 scar covers
electron.. 4 scar covers

PROTON

Fig. 165

193